LABORATORY ASTRONOMY: EXPERIMENTS AND EXERCISES
by Anthony J. Nicastro

ERRATA

Exercise 15, p. 86. The units of Planck's constant should be J-s, not J/s.

Exercise 16, p. 95. The Random Walk is reprinted and adapted from Feynman, Leighton, and Sands, The Feynman Lectures on Physics, copyright 1963 Addison-Wesley, used with permission of Addison-Wesley.

"Experiment" sections of Exercises 10 and 26 quoted and adapted from PH211-212 Laboratory Manuals, used with permission of the Physics Department, Bucknell University.

Exercise 22, p. 139. The attribution for Table 22.3 should read (from H.C. Arp, Astronomical Journal, Vol. 61 (1956): 15-34. Copyright 1956 by the American Astronomical Society.)

Laboratory Astronomy:
Experiments and Exercises

Laboratory Astronomy:
Experiments and Exercises

Anthony J. Nicastro
Bucknell University

 Wm. C. Brown Publishers

Book Team

Editor *Jeffrey L. Hahn*
Developmental Editor *Lynne M. Meyers*
Production Coordinator *Peggy Selle*

Wm. C. Brown Publishers

President *G. Franklin Lewis*
Vice President, Editor-in-Chief *George Wm. Bergquist*
Vice President, Director of Production *Beverly Kolz*
Vice President, National Sales Manager *Bob McLaughlin*
Director of Marketing *Thomas E. Doran*
Marketing Communications Manager *Edward Bartell*
Marketing Manager *David F. Horwitz*
Executive Editor *Edward G. Jaffe*
Production Editorial Manager *Colleen A. Yonda*
Production Editorial Manager *Julie A. Kennedy*
Publishing Services Manager *Karen J. Slaght*
Manager of Visuals and Design *Faye M. Schilling*

Cover image © FourByFive

Cover design by Jeanne Marie Regan

Copyright © 1990 by Wm. C. Brown Publishers. All rights reserved

Library of Congress Catalog Card Number: 88-63834

ISBN 0-697-08475-2

No part of this publication may be reproduced, stored in a retrieval system, or transmitted, in any form or by any means, electronic, mechanical, photocopying, recording, or otherwise, without the prior written permission of the publisher.

Printed in the United States of America by Wm. C. Brown Publishers, 2460 Kerper Boulevard, Dubuque, IA 52001

10 9 8 7 6 5 4 3

Per te, Elvira.

Contents

Preface ix

Exercise		Page
1	Experimental Errors: A Determination of g	1
2	Experimental Errors: Is the Period of a Pendulum Amplitude-Dependent?	6
3	Celestial Coordinates and the Celestial Globe	12
4	Daily and Annual Motions of the Sun	21
5	Resolving Power of the Human Eye	29
6	Optical Properties of Lenses and the Refracting Telescope	31
7	Phases of the Moon	36
8	Graphical Construction of Orbits	40
9	Rotation of the Sun	46
10	Radioactive Decay and the Dating of Rocks	56
11	Rotation of Planets Using the Doppler Effect	59
12	Power Output of the Sun	67
13	Photometry	71
14	Parallax Distance Determinations	78
15	Emission Spectrum of Hydrogen	85
16	Interactions of Matter and Radiation	89
17	Observations of δ Cephei	97
18	Algol and Eclipsing Binaries: Observations	102
19	Algol and Eclipsing Binaries: Simulations	107
20	Stellar Spectra and the Hertzsprung-Russell Diagram	115
21	Distance to Cepheid Variable Stars	125
22	Galactic Distance Determinations Using Novae	131
23	Globular Clusters: Distances and Ages	140
24	Hubble's Law	153
25	Quasar Red Shifts and Distances	160
26	Entropy and Cosmology	165

Appendices
Astrophysical Constants 171
Maps of the Evening Sky 173
Index 187

Preface

Astronomy is a science, and just like other sciences, its foundation rests on experiment and observation. Some experiments to measure astronomically relevant quantities, such as the power output of the sun or the wavelength of the H-alpha line in hydrogen, can be carried out in less than two hours using only a modest amount of equipment. Other important astronomical quantities, such as the value of the Hubble constant, and techniques, such as determining the distances and ages of globular clusters, require observations that are difficult to fit into a two-hour laboratory session. To include crucial astronomical measurements, ideas, and techniques such as these, we need to rely on data obtained by others. This laboratory text is structured to strike a balance between: 1) experiments that require students to make measurements with equipment in order to investigate a phenomenon or idea, 2) exercises that involve the analysis and interpretation of data gathered externally, and 3) observational activities that require an investment in time outside of a scheduled laboratory session and that are at the mercy of weather and finite time constraints.

Because this laboratory text was not developed to be used in conjunction with a particular introductory astronomy text, I have included in each experiment an extensive introduction to the principal idea examined in the particular experiment, an introduction that is intended to develop the idea and fit it into the framework of a modern astronomy course. The introduction to an experiment is, therefore, meant to serve as an orientation for the student. Sometimes within the discourse of an introduction a question is inserted (offset and indicated by a double asterisk, **) that requires a comment or response that need not be written down. These questions provide an interlude and a pause to mull over an idea that should not be passed by but should be considered.

Space for the results of measurements and calculations is allocated in each experiment. Where the number of measured quantities is large and organizing the data is important, tables to be filled in have been provided. Problems addressing important calculations or interpretations of data are interspersed throughout the discussion of an experiment and are denoted by P followed by a number in a running sequence in the experiment. Space for responses to the problems has also been allocated.

An instructor's manual is available to accompany this text. It includes some pedagogical and practical pitfalls that have been encountered over the years during which these laboratory experiments and exercises have been used. Where specialized equipment is needed for an experiment, sources and vendors are listed.

Acknowledgments

Many people contributed to the development of this text, and any weak words of thanks can hardly settle the debt I owe them. I greatly appreciate and gratefully acknowledge the suggestions and refinements offered by the reviewers of this manuscript, Roger A. Bell (University of Maryland), Donald R. Franceschetti (Memphis State University), Laurence W. Frederick (University of Virginia), F. E. Rose (University of Michigan), Charles R. Tolbert (University of Virginia), W. Wolf (S. W. Missouri State University), and especially Karen Kwitter (Williams College). For ideas, inspiration, and constructive dialogue, I owe special thanks to Richard B. Herr, to Michael A. Seeds, and to my colleagues in the Department of Physics at Bucknell.

A. J. Nicastro
Lewisburg, Pennsylvania
November 1988

Exercise 1

Experimental Errors: A Determination of g

Learning Objectives

In this experiment, you will measure the value of the acceleration of gravity g near the earth's surface by timing the oscillations of a pendulum. You will also analyze the data in your experiment to estimate the uncertainty in your value of g.

Introduction

The measurement process is the heart of modern science. Gone are the days when science, or more appropriately, pseudoscience, consisted *only* of philosophical ideas of how the world worked, of models with no basis or recourse to experiments. Certainly, philosophical ideas are at the foundation of many modern theories, but theories are not useful unless they predict and describe accurately the results of experiments. Indeed, a truly useful theory or model of a physical phenomenon correlates the results of experiments and *quantitatively* makes predictions in agreement with experiments. If the predictions of a theory do not agree with the results of an experiment to check the prediction, then:

1. the theory must be modified to account for the results of the experiment;
2. the theory must be abandoned and a new one developed; or
3. the experiment was performed incorrectly and must be retried.

Galileo Galilei di Linceo was a principal framer of the approach of modern experimental science. He insisted that the predictions of models of physical phenomena be tested by experiment, and he demanded that if the model and observations do not agree, then the three options listed above must be explored.

In order to test whether objects of differing mass fell downward with the same acceleration, Galileo performed free-fall experiments and first published comments on them in *De Motu,* written in 1590. There is an oft-repeated and probably accurate account that Galileo made his observations by dropping objects from the Leaning Tower of Pisa, but his most famous comment on the result of his experiments appeared in his *Dialogues Concerning Two New Sciences* (1638), in a literary genre common to his era. He writes:

> Salviati: I greatly doubt that Aristotle ever tested by experiment whether it be true that two stones, one weighing ten times as much as the other, if allowed to fall, at the same instant from a height of, say, 100 cubits, would so differ in speed that when the heavier had reached the ground, the other would not have fallen more than 10 cubits. . . .
> Sagredo: . . . I, who have made the test, can assure you that a cannon ball weighing one or two hundred pounds, or even more, will not reach the ground by as much as a span ahead of a musket ball weighing only half a pound. . . .
> Simplicio: Your discussion is really admirable; yet I do not find it easy to believe that a bird shot falls as swiftly as a cannon ball.
> Salviati: Why not say a grain of sand as rapidly as a grindstone? But, Simplicio, I trust you will not follow the example of many others who divert the discussion from the main intent and fasten upon some statement of mine which lacks a hairbreadth of the truth and, under this hair, hide the fault of another which is as big as a ship's cable. Aristotle says that an iron ball of one hundred pounds falling from a height of one hundred cubits reaches the ground before a one-pound ball has fallen a single cubit. I say that they arrive at the same time. You find, on making the experiment, that the larger outstrips the smaller by two finger breadths . . . now you would not hide behind these two fingers the ninety-nine cubits of Aristotle, nor would you mention my small error and at the same time pass over in silence his very large one.

Notice that Galileo argues that the Aristotelian predictions are certainly in error, but he also realizes that the two balls do not strike the ground at precisely the same instant, i.e., there is an experimental error involved. The relative size of the error, Galileo points out, is small when compared with the distance the balls fell and is still consistent with the idea that the balls fall with the same acceleration.

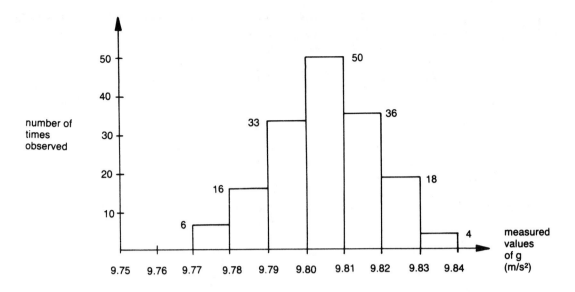

**Galileo eventually went on to attribute the "small error" in the fall times to the effects of air resistance, but without a vacuum pump could not directly substantiate this rationalization. A skeptic might have proposed instead that the differences were due to gravity itself. Comment on what justification, if any, Galileo had for not searching for some formula relating the fall time to the weight of the object.

The principal idea is that *any* measurement is subject to *some* uncertainty, some error, if only because the experimenter did not use equipment of infinite precision. When an experiment is performed or a measurement is repeated, the new value of the measured quantity is probably not exactly the same as the previously measured value. If many determinations are made, the results of the measurements would be expected to all lie near an average value, but there would be a range of measured values around this average.

The results of a long series of experiments to determine g are plotted below. The graph gives the number of times the measured value of g fell within the values indicated on the abscissa. For example, the value of g measured fell within the range 9.80–9.81 m/s² 50 times and within the range 9.83–9.84 m/s² 4 times.

**Verify that 163 measurements of g are represented on the histogram.

From the histogram you can see that most of the measurements of g lie between 9.79 and 9.82 m/s² and it is not surprising that the average value of the 163 measurements is in this range. But notice that the results are spread out over quite a sizable range, even though they cluster near 9.80 m/s².

Examine the following results of two experimenters imaginatively called A and B.

A: 9.81, 9.80, 9.80, 9.79, 9.80   Avg = 9.80

B: 9.60, 9.70, 9.80, 9.90, 10.0   Avg = 9.80

Both experimenters A and B measured values of g and each found that the average of their measurements was 9.80 m/s², yet we would have a little more confidence in Experimenter A's data. Why? The spread in the data for A is much less than the spread in the data for B. If each were to measure again a value for g, we could feel confident that the value measured by A would be closer to 9.80 m/s² than the next value determined by B. Experimenter A's technique or equipment seems to have been better than B's, judging from the narrow spread of A's data.

If A were to report the results of the series of experiments, A might quote a value of g in the form

$$g = 9.80 \text{ m/s}^2 \pm .01 \text{ m/s}^2$$

That is, if the experiment to measure g using A's equipment and technique were repeated, the value obtained would probably be within 0.01 m/s² of 9.80 m/s² most of the time. The 0.01 m/s² represents A's quantitative estimate of the uncertainty in the value of g A has found. It is a measure of the variations that are found (or can be expected to be found) in the values of g. Experimenter B would report the same mean or average value for g, but B would necessarily report a larger uncertainty in the value of g, assuming B is honest.

In order to see a reasonable way to quantitatively estimate the spread in a set of data (which is related to the uncertainty in the measured value), let us answer the question: On the average, how far away from the mean value is any individual value?

In Experimenter B's data, the 9.60 value is 0.20 units away from the mean value of 9.80. It happens to be below 9.80, but we are interested only in the absolute distance from the average value. The 9.70 value is 0.10 units away; the 9.80 value is 0 units away; the 9.90 is 0.10 units away; and the 10.0 is 0.20 units away. The mean (average) of the deviations of B's data from the value of 9.80 is

$$\frac{0.20 + 0.10 + 0 + 0.10 + 0.20}{5} = 0.12 \text{ m/s}^2.$$

We will refer to the uncertainty estimate determined in this way as the mean absolute deviation (MAD).

**Verify that a similar calculation yields a value of 0.004 m/s² for the MAD of A's data.

If we agree to use the MAD as an estimate of the spread in a set of data and as a quantitative measure of the uncertainty in a measured number (there are other ways), then the results of the two experimenters would be reported as

$$A: g = 9.800 \text{ m/s}^2 \pm 0.004 \text{ m/s}^2$$

$$B: g = 9.80 \text{ m/s}^2 \pm 0.12 \text{ m/s}^2$$

One item to note is that the uncertainty in the values for g are small fractions of the value of g itself. For A's data, 0.004 is $0.004/9.8 \approx 0.04\%$ of 9.8 and for B's data 0.12 is 1.2% of 9.8. A and B should count themselves fortunate to have measured a value of a physical constant to an accuracy of about 1%. In astronomy, the physical systems studied are distant and difficult to analyze; uncertainties are higher—usually 10 to 50%. One important constant in astronomy (the Hubble constant) has measured values ranging from 50 to 100 in one set of units!

Suppose that the results of a long series of experiments by many different experimenters indicate that the value of a fictitious quantity Q is

$$Q = 100 \pm 1$$

One experimenter C performs an experiment to measure Q and finds

$$Q_C = 87 \pm 4$$

The subscript C indicates that this value of Q was measured by C. Other experimenters D and E find

$$Q_D = 95 \pm 7$$

$$Q_E = 105 \pm 1$$

**Is the result Q_C consistent with the value of Q above? Is Q_D? Is Q_E?

For those results of experiments that were inconsistent with the value of Q, let us briefly examine a possible cause for the inconsistency. If you judge the *precision* of an experiment by the relative size of the uncertainty, then E has the most precise experiment followed by C and then D. But Q_E is inconsistent with Q because very few, if any, of the different values of Q_E measured by E coincide with the differing value Q measured by the people who determined Q. What might be the cause for such an inconsistency? Suppose E's instruments weren't calibrated properly, e.g., E's meter stick was only 0.9 m long, E's clock gained one minute every hour, or E's one kilogram mass had some chewing gum stuck to the side and actually had a mass of 1.05 kg. You can see that E's measurements would be *systematically* different from someone else's whose instruments had been properly calibrated. In an experiment whose outcome depends on using many pieces of equipment and measuring many different quantities, the result of the experiment can contain systematic errors.

Another way in which one can understand why E's value for Q is inconsistent with the first value is that one of the experimenters did not analyze the sources of error very well. Perhaps one experimenter only measured Q three times and another experimenter measured Q 300 times. Certainly the error estimate in the case where Q is measured many times is better than in the case of only a few measurements.

The Experiment

The period T of a pendulum oscillation is given by

$$T = 2\pi\sqrt{L/g}$$

where L is the length of the pendulum. One period is one complete back and forth oscillation. Carefully measure the length (in meters) of your pendulum. The length is measured from the pivot to the center of the sphere at the end of the pendulum string.

Time 15 oscillations of the pendulum. Displace the pendulum bob so the string makes an angle of no more than 5° with the vertical. You can calculate what maximum distance to the side that you can displace the bob by determining the length of the side X in the triangle that represents a maximally displaced bob.

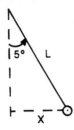

Begin timing the oscillations only after the bob has completed two or so oscillations. (Why?) Determine the period for a single oscillation from the total time elapsed for 15 oscillations. Your measurements can be recorded in table 1.1 below.

Make nine more determinations of the pendulum's period.

Using your ten determinations of the period of your pendulum of length L, obtain ten values of g. Record your ten values of g in table 1.1.

P1. Determine the average value of g for your data.

P2. Determine the mean absolute deviation of your data.

P3. What percentage of your value for g is your MAD?

P4. The value of g is somewhat dependent on the point on the earth's surface above sea level, but a typical value is $g = 9.80$ m/s². Is your determination of g consistent with this value?

P5. Obtain the values of g measured by other lab groups. Is the average of the values obtained by all the lab groups consistent with $g = 9.80$ m/s²?

Table 1.1 Data for a Determination of g

	Time for 15 Oscillations (sec)	Time for One Oscillation (sec)	g m/s²
1			
2			
3			
4			
5			
6			
7			
8			
9			
10			

P6. How could you improve the precision of your measurement of g?

P7. By modifying this experiment, how could you determine if the acceleration of gravity depended on the mass or composition of a falling object?

P8. In 1889, the Hungarian physicist Baron Roland von Eotvos performed an experiment to test whether the gravitational force was precisely proportional to inertial mass. To an accuracy of one part in 10^9 he found that two objects with the same mass interact with the earth with the same gravitational force independent of the material of which they are composed. In other words, Eotvos' answer to the question: Do heavy and light atoms fall with the same acceleration? was 'Yes' to one part in 10^9. When a still more precise experiment was performed in 1964 by Roll, Kratov, and Dicke (checking accelerations toward the sun), the same result was obtained to a precision of one part in 10^{11}. Why do you suppose people went to the trouble of improving the precision of this test? Under what circumstances would it seem important to improve the precision to one part in 10^{13}?

Exercise 2

Experimental Errors: Is the Period Of a Pendulum Amplitude-Dependent?

Learning Objectives

In this experiment, you will measure the period of a single pendulum with two different amplitudes, i.e., sizes of the swing, to determine if the period of the pendulum is amplitude-dependent.

Introduction

Robert Burton (1577–1640) in his book *Anatomy of Melancholy* at one point undertakes a "scientific" discussion concerning the earth's interior:

> Franciscus Ribera will have Hell a material and local fire in the center of the earth, 200 Italian miles in diameter. . . . But Lessius will have this local Hell far less, one Dutch mile in diameter, all filled with fire and brimstone, because, as he there demonstrates, that space cubically multiplied will make a sphere able to hold eight hundred thousand millions of damned bodies (allowing each body six foot square) which will abundantly suffice.

Our ancestors were eager to understand the workings of the world, but had not quite stumbled upon the right approach. Their science was an unbalanced mixture of quantitative ideas, religion, and philosophy. The geology of the earth's core was linked to a two thousand-year-old Greek philosophy!

The person who set science on the right course, to let the results of experiments be a principal factor in the rise and fall of scientific ideas, was Galileo Galilei (1564–1642).

Galileo was the son of an impoverished Florentine nobleman by the name of Vincenzio Galilei. Although Signor Vincenzio was himself much interested in mathematics, he planned for his young son, Galileo, a medical career, which, then as now, was a more profitable profession. Thus, in 1581 at the age of 17, Galileo started the study of medicine at the University of Pisa. Apparently he did not find dissecting dead bodies to be very exciting and his restless mind looked for other kinds of problems.

Galileo became interested in natural philosophy, as physical science was called during the Renaissance, by trying to answer the question posed in the title of this lab. A story, probably apocryphal, tells of Galileo absent-mindedly watching a candelabra while attending mass in the Cathedral of Pisa. The candelabra was set in motion by the attendant who lighted the candles. The consecutive swings were becoming smaller and smaller in amplitude as the candelabra was slowly coming to rest. "Does the time of each swing also become shorter?" Galileo asked himself. Having no mechanical watch—for this had not been invented at that time—Galileo decided to measure the time of the consecutive swings by counting his own pulse. Probably to his surprise, he found that, although the swings were becoming smaller and smaller in amplitude, their duration in time remained the same, at least as far as he could determine by using his pulse. Coming home (and this part of the story is *not* apocryphal), he repeated the experiment with a stone tied to the end of a string and found the same result. He also discovered that, for a given length of string, the oscillation period remained the same no matter whether he used a heavy stone or a light stone in his experiment. In this way, the now familiar device, known as a pendulum came into being. Still having one foot in the medical profession, Galileo reversed the course of his discovery and suggested the use of a pendulum of standard length for measuring the pulse beat of patients. This device, known as the pulsometer, became very popular in contemporary medicine. This was Galileo's last direct contribution to medical science, since the study of the pendulum and other mechanical devices completely changed the direction of his interests and in the way science was pursued.

Perhaps the single most influential idea that Galileo contributed to the methodology of science was the realization that measurements are subject to a variety of indeterminacies that must be involved in the interpretation of any set of measurements. The measurement of a quantity or property, say the period of a pendulum, usually is not precisely reproducible. Measure the period one time, get a number. Measure it again, and the number obtained may be close to the first number, but probably will not be exactly the first number. Measure the period many, many times and what is found is a *distribution* of measurements (see fig. 2.1).

Figure 2.1 Distribution of the results of period measurements of a pendulum.

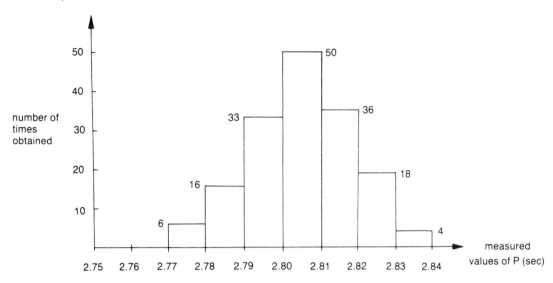

The distribution is characterized by an average value of all the measurements (somewhere between 2.80 and 2.81, judging from the histogram in fig. 2.1) and a width to the distribution that indicates, for this particular experiment, over what range the measured values typically fall.

Let us see how these ideas apply to the situation at hand. We want to answer the question: Does the period of a pendulum depend on its amplitude? How might you check any amplitude dependence experimentally?

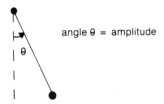

Well, you might measure the time for 15 periods (1 period = 1 back and forth oscillation) of the pendulum at an amplitude of $\theta = 25°$ and divide by 15 to obtain the time for a single period. You would similarly determine the period of the same pendulum with an amplitude of say 5°. You would then subtract the two periods to see if there is a significant non-zero difference.

Note the use of the phrase "significant non-zero difference." Even if we believed the two periods to be the same, surely we do not expect to get *precisely* the same numbers for the periods at $\theta = 25°$ and $\theta = 5°$, so we expect the result of subtracting the two periods to be non-zero. So how do we judge if the expected non-zero answer is significantly different from zero with only one difference determination? The answer is we cannot. We must look at the *spread* of multiple repetitions of the experiment, i.e., we must look at the experimental errors.

A small digression is in order. The term "errors" is quite unfortunate. Most experimental "errors" are not mistakes, as the term "error" might imply. Rather the variations in the results of most measurements are due to the lack of infinitely precise measuring instruments or poor experimental technique. Such "random errors," as they are called, are more commonly encountered than "systematic errors," for example using a clock that runs fast or a 1-meter stick that is only .95 m long.

So, we have to repeat measurements of the period difference between a 25° pendulum and a 5° pendulum and what we will get is a certain spread in measurements about some average value. That spread about the average value we will call the error estimate and will shortly discuss how to quantitatively make the estimate. But for now, let it suffice to say that the results of the period difference will be expressed in the following form:

average value of period difference ± error estimate

Let us discuss three possible results:

A. 0.1 ± 0.5 sec,
B. 0.10 ± 0.09 sec, and
C. 0.10 ± 0.01 sec.

Notice that each of the experimenters obtained the same average value for the period of difference. However, their error estimates differ considerably. For experimenter A, the period differences were typically in the range −0.4 to +0.6. Clearly A's results are *consistent* with there being *no* difference in the periods at different amplitudes. For experimenter C, the period differences were typically in the range +0.09 to +0.11. This result indicates that there *probably* is a significant difference in the two periods. Note that neither A nor C *proved* anything; they only showed that the results of an experiment were or were not consistent with a particular statement. A showed experimentally that the results obtained were consistent with there being no difference, while C demonstrated that the results obtained were consistent with an actual period difference.

Experimenter B's data are a bit more difficult to interpret. B's error estimate indicates that B's determinations of the period differences typically fell in the range +0.01 to +0.19. So, most of the time B obtained a positive result for the period difference, but *some* of the time the period differences obtained may have been out of the quoted range. Suppose we defined the error estimate such that half the time it is in the range and half the time it is out, then you might not feel comfortable saying *anything* about whether the period of the pendulum is amplitude-dependent, and that is perfectly fine.

What are the possible reasons for differing error estimates? A, for example, may have timed the pendulum oscillations using his or her pulse, or a clock that was accurate to only a second. Experimenter A may be a little slow to start and stop the clock or may have timed very few oscillations. Experimenter C, however, may have been using a high-precision electronic clock, perhaps started and stopped electronically when the pendulum bob intercepted a light beam; C's experiment was correspondingly more precise.

You see that regardless of the average value obtained (they all got the same number, viz., 0.1 sec), the **error estimate** was *essential* in interpreting the results of the experiment.

Error Estimates

In this section, we will develop a quantitative way to describe the spread in the values obtained in measuring a quantity in which we are interested.

In order to see a reasonable way to quantitatively estimate the spread, or the variation in a set of data, let us answer the question: On the average, how far away from the mean value is any individual value?

Here is a set of numbers we will use in our discussion:

12, 7, 10, 11, 11, 13, 11, 10, 14
Average value = 11

Look at each individual value and note how far away it is from the average value of 11.

The value 12 is 1 away.
The value 7 is 4 away.
The value 10 is 1 away.
The value 11 is 0 away.   Absolute value of the
The value 11 is 0 away.   deviations away from
The value 13 is 2 away.   the mean value of 11
The value 11 is 0 away.
The value 10 is 1 away.
The value 14 is 3 away.

The sum of the deviations is 12 and so the average deviation is 12/9 = 1.3 (there were 9 measurements); that is, on the average, this experimenter's results vary about 1.3 around the mean value of 11. We will call the error estimate obtained in this way the **mean absolute deviation (MAD)**.

As you can see from the data, most of the values obtained are within 1.3 units of the mean value (6 of the 9 values are within 1.3 of 11), but some of the values were beyond the range.

Why not just quote the extremes of the values obtained as the error estimate, i.e., for our case 11 ± 4 (this would include the "7" measurement)? Stating the answer in this way obscures the fact that most of the measurements were clustered near 11. You could have gotten 7, 7, 7, 7, 11, 15, 15, 15, 15. The average value is still 11, and using the extreme values obtained still give 11 ± 4. But you can see that the results are quite different. In the second case, the results are clustered around *two* different numbers, 7 and 15.

Let us agree to use the mean absolute deviation as an error estimate for multiple repetitions of an experiment. To do this, you must:

1. obtain the mean (average) value for the set of data,
2. take the absolute value of the difference between the mean value and each individual measurement, and
3. obtain the average value of these deviations.

The Experiment

The strategy we will use is to successively measure period differences between a pendulum of amplitude 25° and the same one with an amplitude of 5°. To check the experimental technique, we will perform the experiment again with the amplitudes being 25° in each case; we fully expect the results in the *second* case to be consistent with no period difference (since there is no amplitude difference).

Begin by determining where to place marks on the wall or table near the pendulum bob so as to indicate amplitudes of 5° and 25°. Do not use a protractor to measure the angle. Instead, calculate the distance to the side that you need to displace the bob to obtain the desired angle. That is, determine the distance X along the wall or table you need to produce the angle θ.

(Your data for the experiment will be recorded in the two worksheets.)

Your timing measurements will be made using an electronic timer or stopwatch. When making a timing measurement, allow the pendulum to go through one or two complete swings before beginning to time the oscillations. (Why?) One oscillation is one complete back-and-forth motion.

Time 15 oscillations at 25°. Enter your data on worksheet 1.

Immediately after this, time 15 oscillations at 5° and enter this number in the appropriate space on the worksheet.

Calculate the period for a single oscillation at each amplitude and determine the period difference $T_{25} - T_5$. Your data are recorded on the worksheet.

Repeat this procedure for a total of seven times, giving seven different determinations of $T_{25} - T_5$.

Calculate the average value of $T_{25} - T_5$ and the mean absolute deviation of your individual measurements from the average value of $T_{25} - T_5$. The worksheet will help you organize your results.

Lastly, quote your results in the form

$$T_{25} - T_5 \pm \text{MAD} = \underline{\hspace{2cm}}$$

here and on the worksheet.

P1. Based on your results above, choose *one* of the following:

a. The above result is consistent with there being no difference in the period of a 25° amplitude pendulum and one with a 5° amplitude.
b. The above period is consistent with there being a difference in the period of a 25° pendulum and a 5° one. The period of a pendulum with an amplitude of 25° is (circle one) larger/smaller than with an amplitude of 5°.
c. The above result does not permit us to choose between options a and b above.

Worksheet 1

Name _____

Date _____

Section _____

Trial No.	Time for 15 Oscillations at 25°	T_{25} = Time for One Oscillation at 25°	Time for 15 Oscillations at 5°	T_5 = Time for One Oscillation at 5°	$T_{25} - T_5$
1					
2					
3					
4					
5					
6					
7					

Average value of $T_{25} - T_5$ = _____

Absolute values of the deviations
from the average value of $T_{25} - T_5$.

1. _____
2. _____
3. _____   Mean absolute deviation = _____ (average
4. _____   values of the deviations 1–7 to the left)
5. _____
6. _____
7. _____

Average value of $T_{25} - T_5 \pm$ MAD = _____

10   Exercise 2

Worksheet 2

Name _____

Date _____

Section _____

Trial No.	Time for 15 Oscillations at 25°	T_{25} = Time for One Oscillation at 25°	Time for 15 Oscillations at 25°	T_{25} = Time for One Oscillation at 25°	$T_{25} - T_{25}$
1					
2					
3					
4					
5					
6					
7					

Average value of $T_{25} - T_{25}$ = _____

Absolute values of the deviations
from the average value of $T_{25} - T_{25}$

1. _____
2. _____
3. _____
4. _____   Mean absolute deviation = _____ (average
5. _____   values of the deviations 1–7 to the left)
6. _____
7. _____

Average value of $T_{25} - T_{25} \pm$ MAD = _____

As a control experiment, redo the entire procedure, except that the two amplitudes you will work with will *both* be 25° (see worksheet 2). Certainly, you do not expect there to be a difference between a 25° amplitude pendulum and a 25° amplitude pendulum! It is the same situation!

P2. Are your results in this case consistent with there being no difference between a 25° pendulum and a 25° pendulum? If, in fact, you find no difference, then your experimental technique is sound.

P3. Based on your data, how would you answer the question: Is the period of a pendulum amplitude-dependent?

Exercise 3

Celestial Coordinates and the Celestial Globe

Learning Objectives

This laboratory activity will introduce you to the celestial coordinate system used to locate objects in the sky. You will also use a celestial globe to determine the appearance of the sky various places on the earth at different times of the year.

Introduction

To an observer unencumbered by scientific models, the night sky appears to be an inverted bowl resting on a flat plane. The observer appears to be located at the center of the bowl. Sprinkled over the inner surface of the bowl are the stars arranged in fixed, identifiable patterns that do not change noticeably from day to day, year to year, or even century to century.

Among what ancient observers called the "fixed stars" seven objects move: the sun, the moon, and five apparently star-like objects called planets. Other objects that moved in the sky but were transitory, such as meteors and comets, were considered by the ancient observers, Aristotle in particular, to be atmospheric phenomena and were not considered in modeling the heavens.

The motion of the sun was readily apparent. The general direction of its rising is synonymous with what we call the east. The direction of its setting is the west. Each day the sun rises to its highest altitude above the horizon when it is due south. If we imagine a line extending from the north point on the horizon passing overhead and continuing to the south point on the horizon, then the sun reaches its highest altitude when it crosses this line called the **meridian.** When the sun, or any other object in the sky, crosses the meridian, the object is said to **transit.**

During the course of a single night, the stars wheel around a fixed point as if the celestial bowl on which they are seemingly attached were spinning on an axis passing through this point and the observer's position. On a time exposure photograph as in fig. 3.1 made with a camera pointed toward this fixed point, each star traces out an arc of a perfect circle. Careful observation shows that the bowl rotates from east to west at a uniform rate, completing one revolution in just under 24 hours ($23^h56^m4.^s091$). That is, the time interval between successive meridian transits of a particular star, say, is a bit shorter than the average time between successive meridian transits of the sun, 24 hours exactly. (The word "average" is used because the time interval between successive solar transits varies throughout the year, a complication you can explore in another lab exercise.)

P1. From fig. 3.1, estimate the length of time the camera shutter was left open to obtain the photograph of the star trails.

24 hours

Since the sun rate and the star rate are close but not quite the same, it must be that the sun moves slowly with respect to the background stars, and in fact, the sun's path among the stars can be delineated. By noticing the sun's position among the stars at sunrise or sunset each day, we can plot its path among the stars. This path turns out to be a circle and is called the **ecliptic,** and the sun creeps from *west* to *east* at a nearly uniform rate, taking 365 days and almost 6 hours to complete a single circuit.

Where is this circle on the celestial sphere? In order to answer this question, we will establish a coordinate system *on the sky* so that we can specify locations of points and circles. We will produce an analog of the system of latitude and longitude used to locate positions on the earth's surface.

Figure 3.1 Star trails around the north celestial pole. This photograph is a time exposure taken with a stationary camera. (Lick Observatory Photograph)

The Celestial Pole and Equator

Let us call the fixed point about which the sphere of stars turns the celestial pole. The direction on the ground toward this fixed point is called *north* if we are in the northern hemisphere of the earth. If we are in the southern hemisphere, a different fixed point is apparent, and the direction toward the point is *south*. In the northern sky a bright star, Polaris, is near the north celestial pole. There is no bright star near the south celestial pole.

On the surface of the earth, the line that is everywhere 90° away from the pole is called the equator. It is a circle bisecting the earth halfway between each pole. In the sky, the line (circle) on the celestial sphere that is everywhere 90° from the celestial pole is called the celestial equator. Figure 3.2 shows the geometry of the celestial sphere relative to that of the earth. The angle α formed by the observer, the center of the earth, and a point on the equator is defined to be the observer's latitude on the earth.

P2. Prove that the angle α, the observer's latitude, is in fact the same as the angle α of the north celestial pole over the northern horizon.

Celestial Coordinates and the Celestial Globe   13

Figure 3.2 The geometry of the celestial sphere relative to that of the earth. Angle α is defined to be the latitude of the observer.

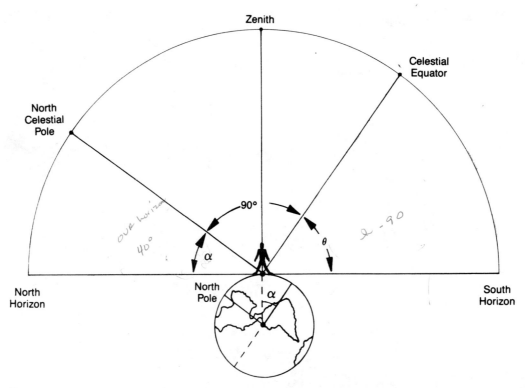

If the celestial pole is an angle α above the northern horizon, then we can establish the location of the celestial equator. From fig. 3.2, the angle along the meridian that the equator makes over the southern horizon is θ = 180° − 90° − α. So the celestial equator starts in the east, climbs up to an angle of 90° − α over the southern horizon, and then drops down to the west point of the horizon.

P3. If the celestial pole makes an angle of 40° over the northern horizon, what angle over the southern horizon does the equator make?

$\theta = 180° - 90° - \alpha = 180° - 90° - 40° = 90° - 40° = 50°$

angle over Southern horizon = 50°

P4. Suppose you are at Quito, Ecuador (latitude = 0°). Where is the celestial equator?

Zenith

The ecliptic is a circle on the celestial sphere that is tipped at an angle of 23 1/2° to the celestial equator. Figure 3.3 illustrates the relationship between the celestial equator and the ecliptic.

Since during the course of a year the sun traces out the path of the ecliptic, it appears at different places with respect to the celestial equator. At one time, the sun is maximally above (north of) the celestial equator, sometimes right on the equator, and at another time maximally below (south of) the equator. Figure 3.4 shows two different orientations of the sun with respect to the celestial equator. The figures are drawn for the case where the north celestial pole lies at an angle of 40° over the north point of the horizon (i.e., latitude = 40° north). The figures are also drawn for noon, when the sun is on the meridian.

14   Exercise 3

Figure 3.3 The ecliptic (dashed line) intersects the celestial equator at two points called equinoxes. The solstices mark the most northerly and southerly points.

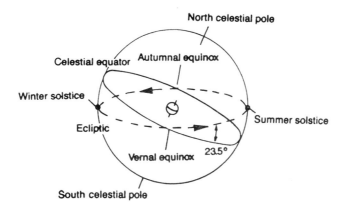

Figure 3.4 a) The diurnal path of the sun across the sky at the summer solstice, b) at the winter solstice.

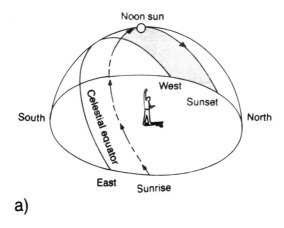

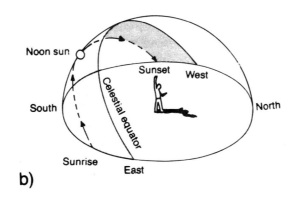

Celestial Coordinates and the Celestial Globe 15

Figure 3.5 The declination of an object on the celestial sphere is its angular distance measured in degrees north (+) or south (−) of the celestial equator.

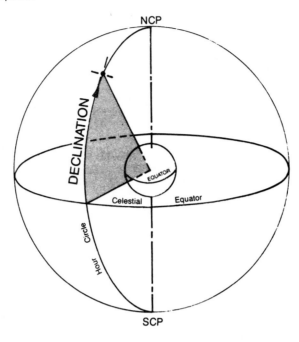

Figure 3.6 The right ascension of an object on the celestial sphere.

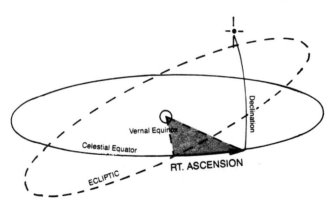

Celestial Coordinates

We can now use the points and circles we have discussed thus far to establish a coordinate system to map the heavens. In analogy with the latitude-longitude system used to specify the location of a point on the earth's surface, we will need two coordinates, called **declination** and **right ascension** to locate a particular point in the sky. The declination of a star is its angular distance measured in degrees between the celestial equator and the point or celestial object. The object's declination is positive (+) if it is north of the celestial equator and negative (−) if it is south. Subdivisions are measured in the usual minutes and seconds of arc. Figure 3.5 illustrates this coordinate. Points on the celestial equator have a declination of 0°; the north celestial pole as dec = +90°; the south pole has dec = −90°.

The right ascension coordinate is a bit more complicated. The right ascension of an object is the angular distance measured along the celestial equator between the vernal equinox and the point on the equator that intersects an hour circle passing through the object (see fig. 3.6). An hour circle is a great circle that passes through both celestial poles as well as through the object itself.

Here is the peculiar aspect of the right ascension coordinate: it is measured in *hours,* with subdivisions of minutes and seconds of *time*. Instead of 360° all the way around the celestial equator, right ascension has 24^h, that is 24^h is equivalent to 360°. Right ascension is measured eastward from the vernal equinox.

P5. How many degrees of arc are equivalent to 1^h of right ascension?

$$\frac{360}{24} = 15°$$

P6. How many minutes of right ascension are equivalent to 1° of arc?

15° ÷ 60 min = .25° of arc per minute   .25 × 4 = 1° arc.

So 4 minutes = 1° arc

P7. How many minutes of arc are equivalent to 1^m of right ascension?

The Celestial Globe

A celestial globe is a representation of the celestial sphere, the stars or other objects in the heavens, together with the coordinate system we have been discussing. Remember, the celestial sphere has you, the observer, at its center, so everything in the celestial globe is plotted from this perspective. As you look at the globe, you will have to visualize that you are at the center of the globe looking out.

P8. Locate on the globe the 1) north and south celestial poles, 2) the celestial equator, 3) the ecliptic, 4) the two equinoxes where the ecliptic and equator intersect, and 5) the solstices.

P9. Which stars are located at the following coordinates?

R.A.	Dec.	Star Name
6^h 43^m	−16° 39′	Sirius
6^h 22^m	−52° 40′	Canopus
18^h 35^m	+38° 44′	Globular star cluster
5^h 12^m	−8° 15′	Rigel
7^h 37^m	+5° 21′	Gemini
14^h 36^m	−60° 38′	Circinus

P10. Which constellation is located at the approximate positions indicated?

R.A.	Dec.	Constellation
11^h	$+50°$	Ursa Major
19^h	$-25°$	Norlsi
3^h	$+20°$	Aries
1^h	$+60°$	Cassiopeia
13^h	$-50°$	Centaurus
7^h	$-40°$	Puppis

The celestial globe is pivoted about the celestial poles and is held in place by a vertical circular ring of metal that is graduated in degrees. A large horizontal ring serves as a representation of the horizon. Adjust the vertical ring so that the celestial pole is at the proper altitude over the horizon for your latitude. (You may need to ask the laboratory instructor for your latitude to the nearest degree.)

As you rotate the globe you will notice that the celestial equator always maintains a single orientation in the sky; it comes up out of the east, reaches a maximum altitude when it crosses the meridian (vertical metal circle), and sinks in the west.

Relative Amounts of Daylight and Darkness

On the ecliptic locate the position of the sun on June 21. Place a small piece of 3M self-stick paper to mark the location. *Do not write on the globe or make any permanent markings.* Note the sun's declination for this day and record it in table 3.1. Now rotate the globe until the sun is in its rising position on the eastern horizon.

**Does the sun rise exactly in the east on this day? no

Rotate the globe until the sun sets and note where on the horizon the sun disappears below the horizon. Bring the sun back to the rising position and by counting how many right ascension hour circles lie between the rising and setting points for the sun, estimate to the nearest 15 minutes the amount of time the sun is above the horizon on June 21. Enter your estimate in table 1. By subtracting this number from 24^h, calculate the length of darkness for this date.

Repeat this for December 21 and March 21.

Readjust the position of the celestial pole to correspond to an observer at Altengaard, Norway (latitude = 70° north). Fill in table 3.2 as you did table 3.1.

Readjust the globe one more time to correspond to an observer in Quito, Ecuador (latitude = 0°). Fill in table 3.3.

P11. *Calculate* the altitude of the sun (in degrees) over the southern horizon at noon on June 21 for an observer at latitude +40° N. Check your answer using the celestial globe.

add horizon to sun   50° + 23.5° = 73.5°

P12. Do P11 for December 21.

Table 3.1 Relative Amounts of Daylight and Darkness for Your Latitude at Different Times of the Year

Date	Dec. of Sun	Number of Daylight Hours	Number of Hours of Darkness
June 21	23.5°		
December 21	−23.5		
March 21			

Table 3.2 Relative Amounts of Daylight and Darkness for Altengaard, Norway (Latitude = 70° N) at Different Times of the Year

Date	Dec. of Sun	Number of Daylight Hours	Number of Hours of Darkness
June 21			
December 21			
March 21			

Table 3.3 Relative Amounts of Daylight and Darkness for Quito, Ecuador (Latitude = 0°) at Different Times of the Year

Date	Dec. of Sun	Number of Daylight Hours	Number of Hours of Darkness
June 21			
December 21			
March 21			

P13. Eratosthenes, who lived in the second century B.C., was a Renaissance man before his time. He was an astronomer, a geographer, a historian, a mathematician, and a poet. With such a diverse background, it is not surprising that he was the director of the Great Library of Alexandria. In one of the scrolls at the library, Eratosthenes read that at noon on June 21 in the southern frontier outpost of Syene, Egypt, near the first cataract of the Nile, obelisks cast no shadows. As noon approached, the shadows of temple columns grew shorter until at noon they were gone. Reportedly, a reflection of the sun could then be seen at the bottom of a well. On June 21 at Syene, the sun was directly overhead. Use the celestial globe to determine the latitude of Syene.

P14. Readjust the celestial globe for your latitude. What range of declinations can stars have so that they never set, that is, are always above the horizon?

Your instructor will provide you with the coordinates of Mercury and Venus for the day of this lab session. Put marked pieces of 3M self-stick paper at the corresponding locations on the celestial globe. Locate the sun for today and place a piece of paper at the sun's location on the globe.

P15. When is Mercury visible in a dark sky? (circle one)

just after sunset   just before sunrise

When is Venus? (circle one)

just after sunset   just before sunrise

P16. Estimate how long each is visible in a dark sky.

Mercury _____

Venus _____

Exercise 4

Daily and Annual Motions of the Sun

Learning Objectives

In this laboratory exercise, you will examine how the daily rising and setting motions of the sun change throughout the year and how the motions of the sun are used in the measurement of time.

☞ **Note:** *A prerequisite for this laboratory exercise is the completion of the exercise "Celestial Coordinates and the Celestial Globe."*

Introduction

Every day the sun rises above the eastern horizon, traces an arched path across the sky, and then sinks below the western horizon. Midway between sunrise and sunset the sun climbs to its highest altitude over the horizon in the south. This daily event, the transit of the sun across the celestial meridian, defines **noon,** a fundamental reference in the measurement of time. The interval from one noon to the next sets the length of the **solar day.**

Subdivisions of the day have had a long and sordid history, and not everyone welcomed the partitioning of the day into smaller units. As Platus lamented c. 200 B.C.:

> The gods confound the man who first found out how to distinguish hours! Confound him too, who in this place set up a sundial to cut and hack my days so wretchedly into small pieces.

The Romans were the principal partitioners of the day. By the end of the fourth century B.C., they formally divided their day into two parts: before midday (*ante meridiem,* A.M., L. before the meridian) and after midday (*post meridiem,* P.M., L. after the meridian). An assistant to the Roman consul was assigned the task of noticing when the sun crossed the meridian and announcing it in the Forum, since lawyers had to appear in the courts before noon.

By the beginning of the Common Era, the Romans eventually made finer subdivisions of the day. The "hours" of their daily lives were one-twelfth of the time of daylight or of darkness. These variant "hours"—equal subdivisions of the total time of daylight or of darkness—were quite elastic and not really chronometric hours at all. For example, at the time of the winter solstice, by our modern measures there would be only 8 hours, 54 minutes of daylight, leaving 15 hours, 6 minutes for darkness. Near the winter solstice, the Roman daylight "hour" corresponds to

$$\frac{1}{12}(8^h 54^m) = 45\ 1/2\ \text{minutes}$$

by modern measure.

**Calculate the length (in modern time units) of a Roman "hour" at night at the time of the winter solstice.

At the summer solstice the times were exactly reversed. One-twelfth of a changing interval of time was not a constant from day to day. These "hours" came to be called "temporary hours" or "temporal hours," for they had meaning and length that was only temporary and did not equal an hour the next day. From the Romans' point of view, both day and night always had precisely 12 hours year round. What a problem for the clockmaker!

Sundials were common and were a universal measure of time. They were handy measuring devices since a simple sundial could be made anywhere by anybody without much in the way of special knowledge or equipment. But the cheery boast "I count only the sunny hours" inscribed on many modern sundials, also announces the obvious limitation of the sundial for measuring time. A sundial measures the position of the sun's shadow; thus, no sun, no shadow. And what do you use at night?

By the Roman era, water clocks became prevalent and served as a way to measure the shadowless and dark hours. Such clocks had a limited precision, at least by modern standards, but we must be amazed *not* that the Romans did not provide a more precise timepiece, but that under their reckoning of hours they were able to provide an instrument that served daily needs at all. It required a hefty amount of ingenuity, but the Romans made their water clocks indicate the shifting length of hours from month to month, rather than from day to day. (The day-to-day changes were too small to be of practical interest.)

Such water clocks were standard features in the Roman courtroom. Usually they took the form of a bowl with a hole in the bottom that when filled to a certain level would empty in a fixed time, usually about twenty of our minutes. A lawyer would assess his case and petition the judge for whatever number of "clepsydrae" he thought would be needed to make the case. Apparently lawyers were no less wordy than they are today. One especially tiresome advocate inspired the Roman wit Martial (c. 40–c. 102) to remark:

> Seven water clocks' allowance you asked for in loud tones, Caecilianus and the judge unwillingly granted them. But you speak much and long, and with back-tilted head swill tepid water out of glass flasks. That you may once and for all sate your oratory and your thirst, we beg you, Caecilianus, now to drink out of the water clock!

The equal hour did not arise until about the fourteenth century. Around 1330 the hour became our modern hour, one of twenty-four equal parts of a day. This new "day" included the night, and it was measured by the *average* time between one noon and the next, the average being over one year. For the first time in history, an "hour" took on a precise, year-round meaning everywhere.

This movement from the seasonal or "temporary" hour to the equal hour is a subtle but profound revolution in human experience. Here was humanity's declaration of independence from the sun, new proof of our growing mastery over ourselves and our surroundings. Only later would it be revealed that we had accomplished this mastery by putting ourselves under the dominion of a machine, with impervious demands all its own.

The Difference between the Solar and Sidereal Day

Why does the sun appear to move in our sky at all? The earth is spinning about an axis once a day *and* revolving around the sun once a year.

The rotation of the earth about its spin axis once a day has the effect that celestial objects seem to spin around the earth once a day—apart from any motion the celestial objects might have relative to the earth. If the earth did not revolve around the sun but simply spun on its axis, the sun would appear to go around the earth once a day—once a sidereal day. But the earth does revolve around the sun and this relative motion introduces a small complication: the time from one noon to the next (one solar day) is not the same as the time for the earth to spin once upon its axis (one sidereal day). This difference is so important that we will examine it in two different ways so that its origin is clear.

First, we will adopt a geocentric perspective, that is, we will consider the earth to be motionless and the celestial sphere and the objects on it to revolve around us. For concreteness, let us pick a fixed point on the celestial sphere, say the vernal equinox (see fig. 4.1), the intersection of the celestial equator and the ecliptic where the sun goes from south of the equator to north of it. (The sun in its motion along the ecliptic is at the vernal equinox on or about March 21.) Let us further suppose that it is about noon on March 21 so that the sun as well as the vernal equinox are on the local celestial meridian as in fig. 4.1a. The celestial sphere rotates (due to, of course, the earth rotating about its spin axis). Sometime during the next day that fixed point on the celestial sphere—the intersection of the ecliptic and celestial equator, the vernal equinox—is again on the meridian. That interval of time is one *sidereal day*. A fixed point on the celestial sphere has gone around once. But in that time the sun has moved eastward along the ecliptic as shown in fig. 4.1b. In fig. 4.1b, it is not quite noon since the sun is not yet on the meridian. We have to wait a bit longer for the celestial sphere to continue to spin to bring the sun up to the meridian.

P1. The sun moves 360° around the ecliptic in 365 days, so the sun moves about 1° per day along the ecliptic. Therefore, the sun is about 1° east of the vernal equinox. Calculate how long you have to wait after the situation depicted in fig. 4.1b so that the sun is on the meridian. Express your answer in minutes.

$$\frac{24h}{x} = \frac{360°}{1°} = \frac{24h \times 60m}{x} = \frac{360°}{1°} = \frac{1440m}{x} = \frac{360°}{1°} = \frac{1440m}{360°} = 4 \text{ min}$$

We have described one way of showing that a solar day is a bit longer than a sidereal day. Now we shall examine the situation from a different perspective, a heliocentric one.

In fig. 4.2, the earth is initially at position A in its orbit around the sun. It is noon for an observer at the foot of the dotted arrow. Consequently, it is midnight for an observer at the foot of the solid arrow at A. One sidereal day later, the earth is at B and the arrows have the same orientation as at A, but at B the daylight dotted arrow does not point to the

Figure 4.1 a) (top) A day in late March with the sun and vernal equinox on the local celestial meridian. b) (bottom) One sidereal day later the earth has rotated once on its axis and the vernal equinox is back on the celestial meridian. However, the sun has moved eastward along the ecliptic.

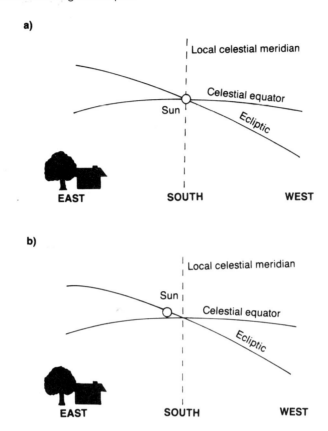

Figure 4.2 The difference between a sidereal day and a solar day arises from the rotation of the earth about its axis and its revolution around the sun. The illustration is a view of the earth and sun from far above the north pole of the earth.

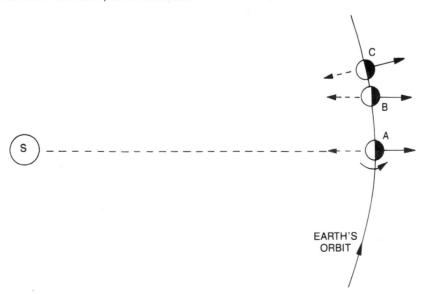

Daily and Annual Motions of the Sun 23

sun and so it is not noon. A little later, the earth has turned more upon its axis and moved a bit in its orbit and now the dotted daylight arrow points toward the sun. The time from A to B constituted one sidereal day, while the time from A to C constituted one solar day.

P2. The diagram below depicts the sun's apparent motion along the ecliptic (where the zodiacal constellations are found) in a heliocentric perspective. As the earth orbits the sun the line of sight of the sun and toward the background stars also moves. For example, the diagram shows the sun in Leo. A month later the earth will have moved enough so that Virgo lies behind the sun.

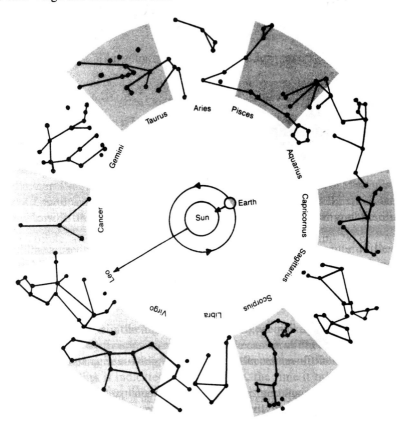

When the sun is in Leo, what constellations will be on the local meridian at midnight? A month later?

a.) aquarius   b) pisces

The Analemma—Details of the Sun's Motion

If the earth's orbit were circular and if the earth's spin axis were perpendicular to the plane of its orbit, the sun would always rise precisely in the east, move along the celestial equator at a constant rate and set precisely in the west. The sun would also appear to move eastward among the stars at a constant rate, completing one revolution in one year. In this idealized situation, the sun would be a perfect clock and would arrive on the observer's meridian at exactly equal intervals.

As you know, the earth's orbit around the sun is elliptical and the earth's spin axis is tilted 23 1/2° from the perpendicular. These circumstances, as you will see, will cause the time interval between successive meridian crossings of the sun to *vary* throughout the year. We can still refer to a fictitious *mean sun* that moves uniformly along the celestial equator and is on the meridian at noon and again precisely 24 hours later. The real sun, unfortunately, does not behave this way, but the corrections to the time kept by the real sun are not large. They can be represented on a three-coordinate plot called the **analemma** (L., sundial).

The analemma is a closed curve resembling a flat-bottomed figure 8 (see fig. 4.3). You may have seen the analemma on a terrestrial globe where it is usually placed in the empty part of the Pacific Ocean. Each point on the analemma

Figure 4.3 The analemma graphs the sun's declination and the daily difference between clock noon and noon by the sundial (sun on meridian) for every day of the year. Declination is the distance in degrees north or south of the celestial equator.

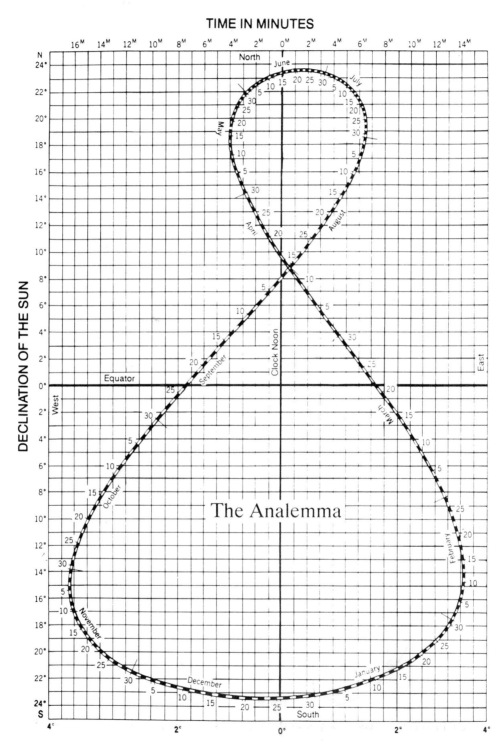

Daily and Annual Motions of the Sun 25

Figure 4.4 Illustration of how the equator-ecliptic angle affects the sun's timekeeping. At the equinox, E represents the solar motion along the ecliptic; its eastward component E' on the equator is shorter. At the solstice, S (equal to E) runs due eastward and the hour circles are closer together, the component S' is longer than both E' and E.

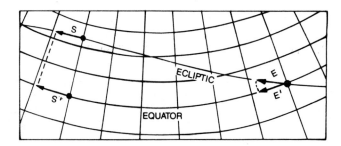

represents a date in the year. The north–south coordinate at the point gives the sun's declination on that date. The east–west coordinate indicates the number of degrees (or minutes of time) by which the sun is east or west of the observer's meridian when the local mean solar time is noon. In the following, we shall examine the origin of the difference between the mean sun and the actual sun, and the information contained in the analemma.

The spin axis of the earth is inclined 23 1/2° to the plane of its orbit around the sun. Because of this tilt, the yearly path of the sun eastward among the stars (the ecliptic) is tilted 23 1/2° with respect to the celestial equator. In late June the sun is 23 1/2° north of the equator and in late December 23 1/2° south of it. This annual north–south oscillation of the sun's declination is responsible for the lengthwise extension of the analemma pattern.

The ecliptic tilt has yet another effect on the sun's motion. Since the sun moves along the ecliptic, which is tilted with respect to the celestial equator, the sun's motion relative to the stars is due east only in late June and late December. Hence, the sun's eastward advance per day is greatest at those times and least in March and September when the ecliptic crosses the equator at a slant (see fig. 4.4).

Because the meridians of right ascension are more closely spaced at declinations of ± 23 1/2° than at the equator, the actual sun's effective eastward motion is faster than that of the mean sun's. Twice a year, near the solstices, the sun arrives later and later on the local meridian because of its relatively fast eastward motion from day to day (look again at fig. 4.1a and b), and as a clock it runs slow. Twice a year near the equinoxes, the sun arrives on the observer's meridian earlier and earlier each day, and as a clock it runs fast. Therefore, two times during the year the actual sun is ahead of clock time and two times during the year it is behind. This effect gives rise to the east–west spread to analemma and determines its general figure-8 shape.

One further influence on the shape of the analemma arises because of the *elliptical* orbit of the earth around the sun. As Johannes Kepler discovered nearly four centuries ago, a planet moves fastest in its orbit near perihelion (point nearest the sun) and slowest at aphelion (point farthest from the sun). Since the earth reaches perihelion on January 3 and aphelion on July 7, the motion of the sun along the ecliptic is faster than average during the winter months and slower than average during the summer months. On January 3, the apparent rate of the sun along the ecliptic is 1.019 degrees per day, while on July 7 the sun moves at a rate of 0.953 degrees per day. The principal effect of this annual velocity variation of the sun is to broaden the southern loop of the analemma and compress the northern loop.

In summary, the analemma graphs the sun's declination, and daily difference between clock noon and noon by the sun (sun on meridian) for every day of the year. Looking at fig. 4.3 we see that the sun is west of the mean sun, that is, ahead of clock noon, from September 1 to December 26, falls behind from December 26 to April 15, then moves ahead again until June 15. It falls behind again until September 1, alternately speeding up and slowing down with respect to clock time.

P3. Briefly describe the effect on the shape of the analemma if the ecliptic-equator angle were to increase.

P4. Obtain from your instructor the latitude of your location and fill it in below.
latitude of your location = 40°
Determine the altitude over the southern horizon of the intersection of the celestial equator and the local celestial meridian.

OURS 50°   NORWAY 20°

P5. Use the analemma in fig. 4.3 and your answer to P4 to fill in the following table. The latitude of Altengaard, Norway is +70°.

Date	Declination of Sun	Noon Altitude of Sun at Your Location	Noon Altitude of Sun at Altengaard
February 15	13° S	37°	7°
June 21	23½° N	73½°	43½°
October 15	8° S	42°	12°
December 22	23½° S	26½°	−3½°

P6. Rip van Winkle awakens from his extended slumber and asks a passerby what year it is so he can determine how long he slept. The passerby quickly responds and then rapidly moves away. (Can you imagine what Rip looked like after twenty years without a shave and a bath, and sleeping under a tree?—the birds, the squirrels—ugh.) Realizing a twenty-year snooze might be a world's record, he thinks he should pin down the date as well as the year. He moves out from under the tree where he slept and measures the altitude of the sun when it crosses the meridian; he finds it is 44° over the southern horizon. He knows that the latitude of his chosen spot on the Hudson River in New York is 42°. On what two possible dates could Rip van Winkle have awakened? He notices that buds are appearing on the tree he slept under. What is the date of his awakening?

A) Oct 6   March 12
B.) March 12

As we saw previously, the analemma graphs the daily difference between clock noon and apparent noon when the sun crosses the meridian. As an example, check the analemma in fig. 4.3 to see that on October 15 the sun will cross the meridian 14 minutes before clock noon.

P7. For the dates given in the table below, use the analemma in fig. 4.3 to determine whether the actual sun will cross the meridian before or after clock noon and by how many minutes.

Date	Sun Crosses Meridian Before or After Clock Noon	Amount of Time Before or After Clock Noon
January 10	after	7½ min
February 15	after	14 min
April 15	after	1 min
July 30	after	6½ min
September 1	after	½ min
October 30	before	16 min

P8. Determine the altitude over the southern horizon of the actual sun and the time when it crosses your local meridian on the dates in the table below.

Date	Altitude of Sun	Time When Sun Crosses Meridian
May 15	18½° N	11:56 AM
June 15	23½° N	12:00 PM
November 4	15° S	11:44:30 AM
December 25	23½° S	11:59:30 AM

Exercise 5: Resolving Power of the Human Eye

Learning Objectives

In this experiment, you will determine the resolving power of the human eye and investigate on what factors it depends.

Introduction

Humans are a very clever lot. In our never-ending attempt to adapt to the world, we sometimes encounter physical obstacles that might hinder the adaptive scheme. We need to fell a tree to build a shelter, but our bare hands are inadequate for the purpose. No matter, we extend the capabilities of our hands and make an axe. The tree is down. We perceive a need to hurl an object a great distance and we invent catapults and ICBM's. We make extensions for our eyes (telescopes) and we see into the great dark between the stars. Our struggle to adapt to the world has usually involved extending the human body and its senses, and in so doing we have given ourselves greater control over our environment and its influences.

But just what can our bodies do? How fast can humans run, how faint a sound can we hear? In this lab experiment, you will examine the resolving power of the human eye; that is, for two point-like objects separated by a particular distance, how far away can you be and still see the two objects as separate and distinguishable?

We will express the resolving power of the human eye in terms of an angle. As shown in the figure below, for two point-like objects a distance d apart, we need to determine the largest distance D from the two objects so that they are still distinguishable. Once d and D are known the angle θ in the figure can be determined. We will refer to θ as the resolving power of the human eye.

When the triangle is long and thin as in the figure below, the angle θ is given to a good approximation by $\theta = d/D$. The angle θ in this relation is in radians. Recall that there are 2π radians in a circle of 360°, so if you want to convert an angle in radians to one in degrees, you need to multiply by $360/2\pi \approx 57.29$.

In science, and often in everyday affairs, a well-phrased question is frequently half the answer. Unfortunately the question "What is the resolving power of the human eye?" is not well-phrased. Does the angle θ depend on the experimental arrangement? If it does, then a unique answer cannot be provided. In this experiment you will investigate the parameters influencing the resolving power and attempt, nonetheless, to answer the central question.

The Experiment

Unlike other experiments you will be doing this semester, you will not be given step-by-step instructions, only some general guidelines and ideas will be offered.

1. Your point-like objects should be separated by about 3–5 mm. The point-like objects themselves should be no larger than about 1/2 mm in diameter; that is, they must be small when compared to their separation.
2. Does it matter that the point-like objects are black dots on white paper or white dots on black paper?
3. Does the resolving power depend on whether the dots are luminous, point-like sources? Does the brightness of the dots matter?

4. Does the color of the dots affect the resolving power?
5. Does it matter whether you start far away and approach the point-like objects or stand nearby and then move away?
6. Do you want to make a single measurement to answer a question, or do you want to make multiple measurements and average?
7. How do your results compare with your partner's and with others'?

Available for your use are various supplies to help you. You may also want to discuss with your instructor other ideas for factors influencing the resolving power of the eye.

P1. Write a short account of what factors influence the resolving power of the human eye and how they influence it.

P2. How would you answer the question: "What is the resolving power of the human eye?"

P3. It is claimed that an eagle can spot a mouse at 3,000 ft ($\approx$1,000 m), although it's not certain just what the eagle is detecting. Estimate the resolving power of the eagle's eye and compare it to your own.

P4. Most people cannot focus their eyes on objects closer than about 10 cm ($\approx$4 inches) from the eye. Using the results of your investigation, what is the smallest object that you can see?

Exercise 6

Optical Properties of Lenses and the Refracting Telescope

Learning Objectives

In this experiment, you will construct a refracting telescope and examine some optical properties of lenses and lens combinations.

Introduction

The earliest type of telescope used to study the heavens was a refracting telescope. In this configuration, an objective lens is used to collect the light from a distant object and focus the light on a plane, forming an image. An eyepiece lens is used to examine and magnify the image formed by the objective (see fig 6.1).

Numerous candidates have been proposed for the honor of the invention of the telescope. Four nations, England, Italy, the Netherlands, and Germany, have been endeavored to secure a decision in favor of one of its own countrymen.

Current evidence favors the Netherlands. The first telescope was probably constructed in 1608 by Hans Lippershey, a native of Wesel and a manufacturer of spectacles in Middleburg. He fabricated his lenses, not of glass, but of rock crystal. A document found in the archives at The Hague shows that on October 2, 1608, Lippershey applied for a patent on an optical instrument that, according to the description, permitted distant objects to be seen distinctly. He was told to modify his construction and make an instrument enabling the observer to see through it with both eyes. This he accomplished the same year. He did not receive his patent, but the government of the United Netherlands paid him 900 gulden for the instrument and an equal sum for two other binocular telescopes completed in 1609.

> ****Lippershey applied for a patent thinking his optical instrument had a practical value. While the patent was not granted, the government of the United Netherlands purchased outright several instruments. What practical value do you think Lippershey and the government of the United Netherlands saw in the telescope?

The use of the new instrument spread over Europe rapidly. In England the mathematician Thomas Harriot had a telescope that magnified objects 50 times, and he observed sunspots and the satellites of Jupiter in 1610. Just prior to Harriot's observations, Galileo Galilei had fabricated his own telescope, searched the heavens with it, and published a book detailing his discoveries.

Galileo was led to construct his own telescope after rumors had reached him regarding the invention of an instrument through which distant objects could be seen clearly. Galileo wrote in his *Sidereus Nuncius,* which was published at the beginning of the year 1610, that he first heard of the invention of the telescope "about ten months ago." He probably heard that the telescope was a combination of concave and convex lenses, and he set to work to devise such an instrument himself. Guided by the hints he received, he soon succeeded. He made a rough telescope with two lenses fixed at the ends of a leaded tube. Both lenses had one side flat; the other side of one lens was concave, and the other lens was convex. The telescope magnified the linear dimensions of objects by a factor of three, making them appear three times nearer. Thereupon, sparing neither expense nor labor, he got so far as to construct an instrument that magnified an object nearly 1,000 times, and brought it more than 30 times nearer.

Galileo went to Venice and showed the telescope to some of the more influential people of the city. Galileo wrote: "Many noblemen and senators, although of great age, mounted the steps of the highest church towers at Venice to watch the ships, which were visible through my glass two hours before they were seen entering the harbor."

Galileo's telescopes were much sought after, and he received numerous orders from scholars, princes, and governments—the Netherlands, the birthplace of the telescope, not excepted.

Galileo turned his telescope toward the moon and discovered mountains and craters; he turned it to Jupiter and saw that it had four satellites that orbited it (January 7, 1610); he pointed it at Saturn and saw the planet as some sort of triple planet—now known to have been due to a low resolution view of the rings; he examined the sun, saw its spots moving, and concluded that the sun rotates. All this was achieved in a flurry of activity in 1610. His observations seemed to have confirmed the Copernican theory, and the cloud of opposition to Galileo began to gather. Some people refused

Figure 6.1 Object, image, and arrangement of lenses in a simple telescope.

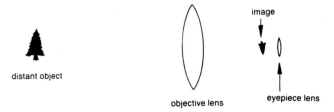

Figure 6.2 Focusing effect of a lens.

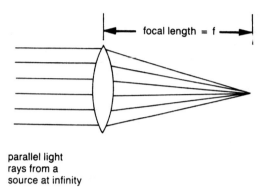

to believe their eyes, and asserted that, while the telescope answered well enough for terrestrial objects, it was false and illusory when pointed at celestial bodies. Others refused to look through it at all. Among the latter was a university professor. Galileo wrote to Johannes Kepler:

> Oh, my dear Kepler, how I wish that we could have one hearty laugh together! Here, at Padua, is the principal professor of philosophy, whom I have repeatedly and urgently requested to look at the moon and planets through my glass, which he pertinaciously refuses to do. Why are you not here? What shouts of laughter we should have at this glorious folly! And to hear the professor of philosophy at Pisa laboring before the Grand Duke with logical arguments, as if with magical incantations to charm the new planets out of the sky.

The antagonism to Galileo and his hated telescope grew stronger. The clergy began to denounce him and his methods. Father Caccini became known as a punster by preaching a sermon from the text, "Ye men of Galilee, why stand ye gazing up into heaven?"

Galileo stood nearly undaunted by withering attacks from both scholars and ecclesiastics and went on to make significant contributions to the study of the motion of objects. The rest you know.

A property of each of the simple lenses that you will be using is that light from a distant object will be focused on a plane on the other side of the lens from the object. The lens, therefore, forms an image of the object at this focal plane. If the object is at infinity, the lens forms the image at a distance f called the **focal length** (see fig. 6.2). An object at nearer distances has its image formed by the lens on a focal plane at a distance greater than f according to the relation

$$\frac{1}{s} + \frac{1}{s'} = \frac{1}{f} \tag{1}$$

where s is the object distance and s' is the image distance. Distances are measured from the center of the lens.

P1. An object is 5 m away from a camera lens, which has a focal length of 50 mm. How far behind the lens is the image formed? This object, a person, moves to within 1 m of the lens. How far must the lens be moved to refocus the image?

Figure 6.3 A ray from the tip of the arrow that emerges parallel to the optical axis gets refracted by the lens so that it passes through the focal point a distance f behind the lenses.

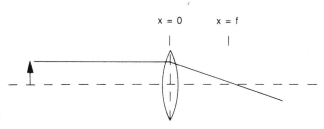

Figure 6.4 Locating the image of the arrow.

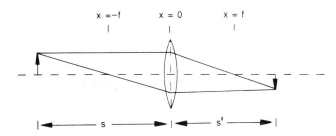

Let us see how the objective lens forms images. A point source of light sends out light rays in all directions. Imagine that an object, an arrow as in fig. 6.3 above, is composed of many point sources of light arranged along its length, but for now, we will concentrate our attention on the source at the tip of the arrow.

One ray will be parallel to the optical axis, and the lens refracts this ray so it passes through the focal point at a distance f behind the lens as in fig. 6.3.

Of course, if light were incident on the lens *from the right* parallel to the optical axis, the lens would refract the ray so that it would pass through the optical axis a distance f *to the left* of the lens. Or conversely, retracing this ray, a ray of light from the tip of the arrow in fig. 6.3 that passed through the focal point *to the left* of the lens would be refracted by the lens so that the ray emerged parallel to the optical axis as shown in fig. 6.4.

The intersection of the two rays in fig. 6.4 indicates the position of the image of the tip of the arrow. Since the bottom of the arrow in fig. 6.4 is on the optical axis the image of the bottom of the arrow must also be on the optical axis.

**On fig 6.4 ray trace the position of the image of the middle of the arrow.

The Experiment

Part 1 Focal Lengths of Lenses

You can get a rough idea of the focal length of the lens by holding the lens horizontally over some white paper and forming the image of some overhead lights. If the overhead lights have diffusers, try forming an image of a window on a piece of paper held vertically. The distance between the lens and the paper is approximately the focal length. (Why approximately?)

For a higher precision determination of the focal length, mount a lens in a holder on the optical bench and point the bench so that the image of some distant object, a far window or, better, some object outside, is formed on a screen positioned behind the lens. Adjust the position of the screen on the optical bench for a sharp image and measure the focal length of the lens.

Compare the focal length determination using the above method with the following autocollimation method. Cover the light source with aluminum foil and make a small hole in the foil with a pin. (The light source should not have a diffuser covering the bulb.) Mount a lens whose focal length you want to determine on the optical bench and place a mirror on the bench so that the lens is between the source and the mirror as shown in fig. 6.5.

Figure 6.5 Autocollimation method for determining the focal length of a lens.

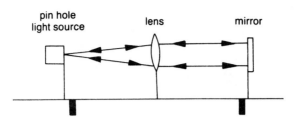

Adjust the position of the lens until a sharp image of the pinhole is formed on the aluminum foil. The distance between the foil and the center of the lens is the focal length. Measure the focal length of each lens in this way.

**Does the position of the mirror matter?

Focal Lengths of Lenses		
Lens Number	Rough Estimate of f	Autocollimation Measurement of f
1		
2		
3		
4		

P2. What is the percentage difference between the focal length as measured by the autocollimation method and the imaging of a distant object? Does one method consistently produce a larger value of f than the other?

Part 2 The Refracting Telescope

Select the lens with the longest focal length and mount it on the optical bench. Place the light source (with no aluminum foil) so that it lies more than about two or three focal lengths in front of the lens. Find the focal plane of the lens by positioning the screen where the image of the illuminator is sharpest. Measure the object distance and the image distance. Verify that the relation in eq. 1 is correct by determining f using s and s', and by comparing f obtained in this way with the other methods. Note the relative size and orientation of the image.

To examine the image formed by this **objective lens** (it points to the object), a high power (short focal length) lens is used. Remove the screen and mount one of your other lenses on the optical bench and view the image formed by the objective with this **eyepiece lens.**

P3. Is the image now magnified or diminished?

P4. Replace the eyepiece lens with other lenses. How does the image size depend on the focal length of the eyepiece lens?

Part 3 Light-Gathering Power

As you have seen, a lens collects the light incident on the surface and forms an image. A light source radiates electromagnetic waves in many directions, but, of course, only the light that actually strikes the lens can go to form an image; the other radiation just continues on its way.

When you stand beneath the canopy of stars on a clear evening, light from any one particular star is streaming down all around you, but only the light that actually enters your eye contributes to the perception of that star. What a waste, all that radiation just dribbling on the ground! Of course, more than just what your eye lens collects can be utilized—use a bigger collector, a telescope lens.

Collecting visible light and other forms of electromagnetic radiation is actually the principal function of an astronomical telescope. The objects in the astronomical universe, even though they may be powerful energy sources, are so incredibly distant that their energy is vastly diluted by the time it reaches us on earth—stars, nebulae, and galaxies, by and large, are *faint*. To detect these things—to *see* them—as much radiation is collected as is practical.

In this next portion of the experiment, you will investigate how the amount of light collected by a lens depends on the size of the lens.

Measure the "size" of each of the lenses you have. We will characterize the size of a lens by its cross-sectional area, or equivalently, its diameter or radius.

Choose the largest lens and hold it so that the image of a nearby light bulb is formed on the photosensitive surface of a photoelectric detector. Record the maximum output of the detector; you may have to move the lens around a bit to find the maximum. The photoelectric detector produces a current (or voltage) that is proportional to the amount of light striking the photosensitive surface. Continuing with the other lenses, being careful to maintain the same lens-light source distance, record the maximum output of the detector.

Light-Gathering Power Lenses			
Lens Number	Diameter of Lens	Detector Output	Predicted Value of Constant or Detector Output
1			
2			
3			
4			

Check to see that the detector output is directly proportional to the *square* of the diameter of each lens, that is, the relation is one of the form

$$\text{detector output} = \text{const} \times d^2$$

Determine the constant for the first lens and see if measurements with other lenses are consistent with the above relation. You can do this by:

1. determining the constants for all the other lenses and checking to see if they are in some sense consistent, or
2. using the constant you determined for the first lens to see if you can "predict" the detector output for lenses of other sizes.

Optical Properties of Lenses and the Refracting Telescope

Exercise 7

Phases of the Moon

Learning Objectives

Using about six observations of the moon over a two-week interval, you will examine how the moon changes position in the sky and how its appearance changes as it moves.

Introduction

Around the world and in every era, people have scrutinized the moon. Its influence has seeped into our language, where we find relics of mythic, mystic, and romantic meanings—in such words as "moonstruck" and "lunatic" (Latin *luna* means moon), in "moonshine," and in the moonlit setting of lovers' meetings. Even deeper is the primeval connection between the moon and measurement. The word "moon" in English and its cognates in other languages are rooted in the base *me* meaning measure, as in the Greek *metron*, and in the English *meter* and *measure*, reminding us of the moon's service as the first universal measure of time. The ancient German communities, Tacitus reported nearly two thousand years ago, held their meetings at new or full moon, "the seasons most auspicious for beginning business."

The phases of the moon were convenient worldwide cycles that anybody could see and so were the basis for reckoning time in the construction of a calendar. The Babylonians and ancient Israel, Greece, and Rome all used lunar calendars. The Muslim world, with a literal obedience to the words of the prophet Muhammad and to the dictates of the holy Koran, continues to live by the cycles of the moon.

Despite or because of its easy use as a measure of time, the moon proved to be a trap for naive mankind. For while the phases of the moon formed a convenient cycle, they were an attractive dead end. What hunters and farmers most needed was a calendar based on the seasons—a way to predict the coming of rain or snow, of heat and cold. How long until planting time? When to expect the first frost? The heavy rains?

For these needs the moon gave little help. The seasons of the year, as you know, are governed by the movements of the earth around the sun or equivalently, the movement of the sun against the background stars. The discomforting fact that the cycles of the moon and the cycles of the sun are incommensurate would stimulate thinking, however. Had it been possible to calculate the year, the cycle of seasons, simply by multiplying the cycles of the moon, humans would have been saved a lot of trouble. But we might also have lacked the incentive to study the heavens and to become astronomers.

The Observations

A complete cycle of lunar phases requires about 29 1/2 days. You will follow the moon for at least half a cycle so your observations will be spread over the course of about two weeks. During this time you will be recording both the appearance of the moon, that is, how much of the lunar disk is illuminated, and also the *position* of the moon relative to the landscape and the background stars.

A. Choosing a Location for Your Observations

You will need to find a spot with a relatively unobstructed view of the horizon and the sky. The middle of a large open space with a few low-profile landmarks near the horizon is ideal. You will need a little ambient light to make some sketches to record your observations, but you should not be near any bright sources of light. *It is important that you use precisely the same location for all your observations,* so if you make them from the middle of an open field or a parking lot, mark the location of your observing spot so that you can return to it.

B. Timing Your Observations

Your instructor will give you some general guidelines as to dates and times for when to begin your series of observations, but you can choose a time of night convenient to you for observing the moon. Once you choose a time of night for observing, *it is critical that you make all your observations at the same time of night to within 15 minutes.*

Because the weather may hamper your observations over the course of the observing period, you do not want to let many clear nights go by. You should try to make observations every other night, weather permitting. After your first observation, which might require about half an hour, succeeding observations of the moon will require much less time, about five minutes.

C. Recording the Observations

In order to record the appearance and position of the moon, you will need to make a scale drawing of the moon in its relation to the landmarks around your observing site. This sketch can be made on a standard poster board (22″ × 28″) or a similarly sized sheet of drawing paper. You may want to initially go out about a half hour before the optimum observing time that you have chosen so that you can get enough information to complete the sketch prior to your first lunar observation.

To make a scale drawing, you will need to measure the height, width, and separations of a few landmarks around your observing site. One way is to make measurements in degrees. Convenient standards are the following:

1. the width of your outstretched hand held at arm's length is about 22°
2. the width of your fist held at arm's length (knuckle to knuckle) is about 8°
3. the width of your three middle fingers held at arm's length is about 5°
4. the width of your index finger at arm's length is about 2°

A convenient scale to use would be 5° in the sky and landscape corresponding to 2 centimeters on the drawing.

Alternately you can carry a ruler with you and measure the dimensions and separations of landmarks with the ruler held at arm's length. In this case, a convenient scale to use is 2 centimeters on the ruler corresponding to 1 centimeter on the drawing.

To plot the position of the moon on your drawing, you will need to make two measurements. Choose a convenient landmark near the moon and from a particular point on the landmark measure *how far over* and *how far up* the moon's center is located.

**Estimate the angular size of the moon. (The width of your index finger at arm's length near the top knuckle is about 2°.)

Record the appearance of the moon at the plotted position. You may want to forget about getting the moon's size to scale and use any conveniently sized object like a coin to draw in the moon and indicate the degree to which it is illuminated.

Record the positions of a few bright stars near the moon on your first observing session.

Number your observations of the moon and in a corner of your drawing record the date and time to the nearest minute of each observation.

Determine the direction of one or two compass points (e.g., west and south) and indicate them on your drawing.

P1. How would you describe the night-to-night path of the moon with respect to the landmarks around your observing site?

P2. In which direction does the moon move with respect to the background stars from night to night?

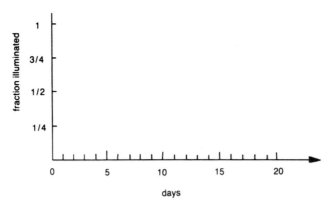

On the graph above, plot the fraction of the lunar disk illuminated as a function of the number of days after your first observation. (Your first observation is at 0 days.)

For each of the moon's positions 1 through 8 in fig. 7.1, shade the dark portion of the moon that would be seen for an observer high above the plane of the earth-moon system.

In fig. 7.2, shade in below what an earth-bound observer would see for the corresponding positions.

Lunar Phase Nomenclature

The explanation of the phases is quite simple: the moon is a sphere and at all times the side that faces the sun is illuminated and the side that faces away from the sun is dark. From the previous drawings, you can see that the phase of the moon that we see from earth, as the moon revolves around us, depends on the relative orientation of the sun, moon, and earth.

When the moon is almost exactly between the earth and sun, the dark side of the moon faces us. We call this "new moon." A few days earlier or later we see a sliver of the lighted side of the moon and call this a "crescent." As the month wears on, the crescent gets bigger until half the disk of the moon is illuminated. We sometimes call this "half moon." However, since this occurs one-fourth of the way through the cycle of phases, the situation is also called a "first-quarter moon." (Certainly, someone owes you an apology for calling the same phase both "quarter" and "half," but since I do not know who to blame, we will continue on and pretend it does not bother us.)

When over half of the moon's disk is visible, we have a "gibbous" phase. When the moon is on the opposite side of the earth from the sun, the entire face of the moon is illuminated; this is a "full moon." A few days later the moon is noticeably in a gibbous phase again and soon thereafter we see a half moon again, called third-quarter this time. Then we go back to a crescent phase and "new moon" again. The cycle then repeats.

When the amount of illuminated moon is increasing from day to day, the moon is said to be "waxing." So from new moon to full we have waxing crescent and gibbous phases. When it is decreasing, after full moon until the next new moon, the phases are said to be "waning."

P3. In the spaces below, write down the names of the phases corresponding to sketches in fig. 7.2. Include whether a phase is waxing or waning.

1. _____   5. _____
2. _____   6. _____
3. _____   7. _____
4. _____   8. _____

Positions a, b, c, and d on fig. 7.1 correspond to observers looking at different times of the day. Observer a is observing halfway between sunrise and sunset, i.e., at noon. Observer b is on the boundary between day and night and so is observing at sunset. (Why not sunrise?) Observer c is observing at midnight, while d is at sunrise.

P4. Place a cross on the earth representing your observing time.

Figure 7.1 The earth–moon system as viewed from far above the earth's north pole with different moon positions 1–8.

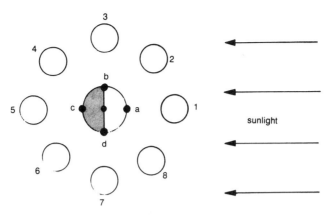

Figure 7.2 Lunar phases seen by an earth-bound observer.

P5. At what time of the day do you expect:

a) the full moon to rise?

b) the new moon (or a very thin crescent moon) to rise?

c) the first-quarter moon to rise?

d) the third-quarter moon to rise?

e) the first-quarter moon to set?

f) the new moon (or a very thin crescent moon) to set?

Exercise 8: Graphical Construction of Orbits

Learning Objectives

By graphically modelling the action of a large mass on a much smaller mass, the dynamics of the solar sytsem will be explored.

Introduction

How can we describe and predict the motion of objects? What influences the motion of objects? The search for answers to these questions has a long and venerable history. Prior to the seventeenth century, the key to understanding the motion of objects was the idea of **natural place**. In the Aristotelian picture, each of four elements—Earth, Water, Air, and Fire—which were supposed to be constituents of all things, has a natural place. When in that place, it will remain at rest; removed to a distance, it will tend to move back. The natural place of the element Earth is at the center of the universe (which happened to coincide with the center of the earth—that is how the earth got there). Thus, Earth falls when released from the hand.

Water also tends to fall because its natural place is a spherical shell about the center of the earth. On the other hand, Fire and Air tend to rise because their natural places are spherical shells whose surfaces were above the surface of our world. However, because the four elements are continually interfering with each other, they are prevented from separating into their natural spheres. Thus, the Water in a crucible does not fall because it is constrained by the Earth forming the crucible.

Aristotle explained the fall of a stone in the following way: when we pick up a stone (whose major constituent is the element Earth), we move it farther away from its natural place, the center of the universe. When we release it, the stone strives to reach its natural place, moving in a *straight line* toward the center of the universe. The potentiality of falling, inherent in the stone, becomes actuality.

One might question whether this is an explanation at all. It seems tautological to say that a stone falls because Earth has a tendency to fall. Would not a satisfactory explanation necessarily involve some *mechanism,* some sort of invisible strings, magnets, or vortices that push or pull the stone downward? Some sort of *force?*

The horizontal motion of a stone tossed some distance away presented a particular problem. The vertical motion was explained by appealing to the idea of natural place, but what explains the sideways motion?

Progress on this question did not come until the fourteenth century, with the introduction of the idea of **impetus.** The tossed stone was imagined to receive from the person throwing it a quantity of impetus, which somehow kept it going during its flight until the impetus ran out and the projectile fell to earth.

> In the stone or other projectile there is impressed something which is the motive force of that projectile. . . . The projector impresses in a moving body a certain impetus. . . . It is by that impetus that the stone is moved after the projector ceases to move.
>
> —Jean Buridon
> *Questions on the Eight Books of the Physics of Aristotle* (1509)

The details of the various theories of impetus are not important. It suffices to note that all were attempts to answer what was a fundamental question in the context of the Aristotelian world view: For motion like that of a projectile, what is the mover, the source of the motion?

The final break with Aristotelian physics came when Galileo Galilei considered horizontal motion, i.e., motion parallel to the surface of the earth. As we have seen, the Aristotelian view was that such motion required something to sustain it. Galileo suggested an almost diametrically opposed interpretation. In the absence of an external influence such as friction, the object will not slow, Galileo asserted, but will continue to move with a constant speed. Let us examine his argument from *Dialogues Concerning the Two Chief World Systems:*

> Suppose you have a plane surface as smooth as a mirror and made of some hard material like steel. This is not parallel to the horizon but somewhat reclined and upon it you have placed a ball which is perfectly spherical and of some hard and heavy material like bronze. . . . On the downward inclined plane, the heavy moving body spontaneously descends and continually accelerates, and to keep it at rest requires the use of a force.

Now he imagined the plane tilted upward:

> On the upward slope, force is needed to thrust it (the ball) along or even to hold it still, and motion which is impressed upon it continually diminishes until it is entirely annihilated. . . . Now . . . what would happen to the same movable body placed upon a surface with no slope upward or downward(?) . . . There being no downward slope, there can be no natural tendency toward motion; and there being no upward slope, there can be no resistance to being moved, so there would be an indifference between the propensity and the resistance to motion.

Thus, he concluded the ball on the level plane will either remain at rest if initially placed at rest, or

> . . . if given an impetus in any direction . . . it would move in that direction. But with what sort of movement? One continually accelerated, as on the downward plane, or increasingly retarded as on the upward one?

"Neither," he answers; on the level plane "I cannot see any cause for acceleration or deceleration, there being no slope upward or downward." Consequently, if the ball were placed initially in motion, it would continue to move with constant speed "as far as the extension of the surface continued without rising or falling."

This principle is involved in the motion of the planets about the sun. The natural motion of an object is to continue moving at constant speed along a straight line, a single direction.

But the planets do not follow a straight-line path. Something must be diverting the planets from a straight path; some force must be involved. The operation of the force, called **gravity,** was investigated by Sir Isaac Newton.

Prior to examining Newton's Law of Gravitation and the orbital motion of planets, we will build a quantitative basis for describing motion.

Imagine an object moving along a straight line with a constant speed, say 10 m/s. In any one-second time interval, this object will cover 10 m of distance. From some starting point, we can determine how far away the object will be at any time by the relation

$$x = v_0 t$$

where x is the distance from the starting point and t is the time the object has been moving with velocity v_0. If we were to plot the velocity as a function of time, the graph would look like

If we watched this object for 5 seconds and it was traveling with $v_0 = 10$ m/s, the graph would look like:

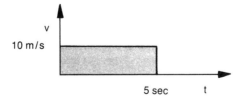

The area of the rectangle (area = [(10 m/s)(5s) = 50 m] is precisely the distance traveled by the object in 5 sec.

Suppose now that the object was accelerated, that its speed was changing. Suppose that after each second, the velocity of the object increased by 2 m/s. If the object started out traveling at 10 m/s, after one second it would be at a speed of 12 m/s. After 2 seconds, the speed would be 14 m/s, and so on. That is, the velocity at any time would be

$$v(t) = 10 \text{ m/s} + (2 \text{ m/s each sec})(t).$$

The rate change in speed of 2 m/s each second (2 m/s²) is called the acceleration of the object. In general,

$$v(t) = v_o + at,$$

in which v_o is the initial speed.

How far does an object go if it is accelerating? Look again at a velocity vs. time graph.

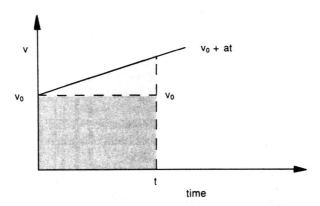

The distance traveled after a time t would be the area under the solid v vs. t line. That portion of the area shaded by lines represents the distance the object would have traveled had it *not* been accelerating. If it simply traveled at constant speed, that distance is $v_o t$ as before. But because of the acceleration, because of the fact that the speed had been *increasing* with time, the distance traveled is greater than the case with constant speed. This extra distance traveled is represented by the area shaded by dots. That triangular portion has area 1/2(base)(height) = 1/2 (t)(at) = 1/2 at². Consequently, the total distance traveled by this accelerating object is

$$x = v_o t + 1/2\, at^2.$$

If the object does not start at $x = 0$, but some other point x_o,

$$x = x_o + v_o t + 1/2\, at^2.$$

**An object is dropped ($v_o = 0$) from the top of a 30 m building. How long does it take to fall? The acceleration of gravity near the earth is 9.8 m/s².

The positions and velocities we have been discussing have an additional character that we haven't explored yet: their directional character. It is not complete to simply say that starting from a point p an object moved x meters. The final position of the object is not known since the direction the object moved was not specified.

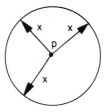

The object could be located anywhere along a circle x meters in radius and centered on the starting point. However, as soon as a direction is specified, all ambiguity is removed.

The idea of velocity also has a directional character associated with it. To start in Tulsa, Oklahoma and travel at 55 miles/hr, trying to get to San Francisco, California, the voyager had better adopt a speed of 55 mi/hr west if the traveler wishes to be reasonable; traveling 55 mi/hr east would take considerably longer.

We will keep track of quantities that have some sort of direction associated with them by writing little arrows over symbols representing directional quantities. Thus, moving a specified distance in a specified direction will be represented by x, an example of which is 10 m east. Velocity, specified by v, might have as an example: 5 meters/sec straight up. So, the relations we obtained previously can be better written

$$\vec{v}(t) = \vec{v}_o + \vec{a}t$$

and

$$\vec{x} = \vec{x}_o + \vec{v}_o t + 1/2\, \vec{a}t^2.$$

The Experiment

The solar system's dynamics are primarily determined by the gravitational attraction of the sun on the planets. In this lab, we model graphically the action of a central force (i.e., a force directed toward a fixed central point) on an orbiting body.

Newton's universal law of gravitation states that the magnitude of the gravitational force between any two bodies is given by

$$F = \frac{GMm}{r^2}$$

and is directed along the line between the bodies. Here G is the constant of gravitation, M and m are the bodies' masses, and r is the distance between their centers. Applying Newton's second law F = ma to the body of mass m, we find its acceleration law

$$a = \frac{GM}{r^2},$$

and is directed along the line from M to m.

By choosing the units properly, we can make the orbits fit on paper and still have an accurately scaled representation of possible planetary orbits. In our case, GM = 8,000, m = 1, and r is measured in centimeters.

Method

We start by fixing a center to represent the sun. Then we choose an initial position and velocity. The acceleration law determines the velocity changes, and the velocity determines how the position changes. If we take small enough time intervals, we can treat the acceleration as constant during an interval. After n time intervals, the equations we need are then

$\vec{a}_n = -\hat{r}_n' (8{,}000/r_n'^2)$ ($\hat{r}_n'$ is a vector of magnitude 1 and is directed along the line from M to m)

$\vec{v}_{n+1} = \vec{v}_n + \vec{a}_n(\Delta t)$

$\vec{r}_{n+1} = \vec{r}_n + \vec{v}_n(\Delta t) + 1/2\, \vec{a}_n(\Delta t)^2$

To start, set up a table of values for n, t, $v_n\Delta t$, r_n', a_n, and $a_n(\Delta t)^2$. (Recall that symbols without vector signs represent magnitudes only.) Then follow the steps below, beginning with n = 0.

Step 1. Locate the initial position $\vec{r}_n$ and plot the vector $\vec{v}_n\Delta t$. This is how much the planet would have moved with no acceleration. See fig. 8.1.

Step 2. Measure the distance r_n' from this point to the center. Use this distance in the acceleration law. See fig. 8.1.

Step 3. Plot the vector $\vec{a}(\Delta t)^2$ directed toward the center. The third side of this triangle will be $\vec{v}_{n+1}\Delta t$. This determines the direction of new velocity $\vec{v}_{n+1}$. See fig. 8.2.

Step 4. The midpoint of the vector $a(\Delta t)^2$ represents the new location of the planet, $\vec{r}_{n+1} = \vec{r}_n + \vec{v}_n\Delta t + 1/2\, \vec{a}(\Delta t)^2$.

Step 5. Move the vector $\vec{v}_{n+1}\Delta t$ so that its *tail* is at the point found in step 4. This can be done by extending the vector (in light pencil) and then moving your ruler so that it is parallel (eyeball is okay) and passes through r_{n+1}. See fig. 8.3.

Now go back to step 2 and continue.

Discussion

As you continue the process described above, you will be tracing out the orbit around the center of attraction. Depending on your starting conditions, the speed may be increasing or decreasing; likewise for the distance from the force center. You should be able to explain qualitatively when the speed will be decreasing and when it will be increasing. Where would you expect the maximum speed? The minimum?

Figure 8.1

Figure 8.2

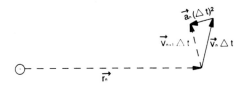

Figure 8.3

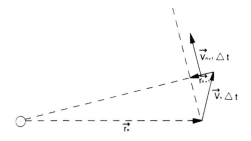

Numerical Errors

Unfortunately, any numerical method that tries to approximate calculus will introduce errors into a calculation of this kind. Unless we are really willing to let Δt go to zero and repeat the steps an infinite number of times, some errors will remain. The source of error in this calculation is our assumption that the acceleration is constant over the interval. We actually used the acceleration near the *end* of the interval. How could we reduce error without making Δt smaller?

Getting Started

A conveniently sized orbit may be sketched in a reasonable number of time steps if the following values are taken:

$$r_0 = 20 \text{ cm}$$
$$v_0 \text{ from 19 to 21 cm/s}$$
$$\theta_0 \text{ from } 90° \text{ to } 135°$$
$$\Delta t = .2 \text{ s}$$

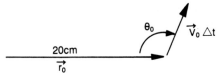

Exercise 8

The period of the orbit can be calculated analytically from the starting values and the acceleration law. A table appears below relating initial velocity to the number of time steps ($\Delta t = .2$ s) needed to complete a closed orbit.

v_o	# of Steps
19	27
19.3	28
19.6	30
20	31
20.3	33
20.6	35
21	37
22	45
24	75
26	180
28	3927

You should continue for at least half an orbit, about 15–25 steps. Stop if the radius becomes smaller than about 10 cm, since the acceleration becomes too large here.

After trying an orbit from the given starting ranges, take initial conditions and time steps of your own choosing. See what happens!

Exercise 9

Rotation of the Sun

Learning Objectives

Using observations of sunspots made by Thomas Herriot in the year 1612, you will determine the period of rotation of the sun about its axis.

Introduction

Up until the 1600s, the prevalent model of the universe had the earth at its center and all other celestial objects revolving around it. This model had its foundation in Aristotelian physics, which explained the motions of objects in the universe in terms of a curious but workable amalgam of philosophy and observations. In the Aristotelian model the physical universe was divided into two realms: the transitory, corruptible realm extending from the earth's center to just inside the sphere of the moon's orbit, and the external realm beginning at the sphere of the moon and extending beyond. The celestial objects in the external realm followed different rules from objects on earth and were changeless, incorruptible, and eternal.

This system of the universe changed suddenly and violently in the 1600s, but the roots of the upheaval lie with Nicolas Copernicus and the publication of an imprecisely reasoned book in 1543. In the book *Concerning the Revolutions of the Celestial Spheres* Copernicus posited a model of the universe in which the earth was removed from the unique position at the center of the universe and was treated as a planet that, like other planets in his model, circled the sun. The sun was now at the center of the cosmos.

The Copernican view was not adopted immediately and, in fact, was violently opposed for over sixty years. The system had many problems and inconsistencies that troubled the scientists of the time, in addition to theological implications that troubled the clerics. One of the substantive objections to the system dealt with the sun. Copernicus in his book asserted that the earth and other celestial objects rotate about some axis because they have a spherical shape. Copernicus argued that it is the "nature" of a sphere to be in "spherical motion." The problem internal to the model was that the sun, which has a spherical shape, stood still in his system. This inconsistency puzzled readers of his book and represented a serious flaw.

The Aristotelian world system that Copernicus shook was finally uprooted, however, by the observations made soon after the invention of the telescope. Observations with the newly invented telescope removed opposition to a sun-centered system from any claim to scientific respectability and reduced it to mere stubborn obscurantism. The first users of telescopes in the early 1600s must have appeared to themselves as blind men given back their vision. They could see, literally and figuratively, that a sun-centered system was a good model.

Four people figure prominently in early astronomical advances with telescopes: Thomas Harriot (1560–1621) in England, Galileo Galilei (1564–1642) in Italy, Johann Goldsmid (1567–1616) in Holland, and Christopher Scheiner (1575–1650) in Germany. When telescopes were turned toward Jupiter, it was discovered that the planet had four satellites that circled it steadily, thus disproving once and for all that earth was the center about which *all* things turned. They found that Venus showed a full cycle of moon-like phases as the Copernican model predicted it would have and where earlier models predicted otherwise. Through the telescope, they also saw that the moon had a landscape covered by mountains, craters, and what they took to be seas, showing that it (and by extension the other planets) was a world like the earth and, therefore, presumably subject to the same laws of change, corruption, and decay. And when they looked at the sun, they saw that the surface had spots.

To the early telescopic astronomers, a sun with spots was a deep wound to the Aristotelian view of the cosmos. Galileo in 1613 published the book *History and Evidences about the Spots of the Sun* where for the first time in writing he proclaimed his support for the sun-centered system. The Italian word rendered as "spots" in this instance is better translated as "blemishes" and herein lies the source for the avid interest shown this phenomenon. The old view was that the sun was composed of *pure fire,* and the emphasis was fully as much on the first word as on the second. Consequently, people regarded the existence of spots as impeaching the pure nature of the sun, because spots (or blemishes) and purity could not be reconciled. Furthermore, Harriot, Galileo and Goldsmid (who is generally known by his Latinized name Fabricius) inferred from observations of sunspots that the sun must rotate, an apparent vindication of yet another Copernican prediction. With the sun no longer perfect, doubts gradually grew as to its perpetuity, too. Science now began to inquire about the source of the sun's energy (with answers not forthcoming, however, until 1938). Solar physics began here in the first decade of the seventeenth century.

Thomas Harriot—A Brief Perspective

Though recognized by his contemporaries as England's most profound mathematician, most imaginative and methodical experimental scientist, the first of all Englishmen to make a telescope and turn it toward the heavens, Thomas Harriot had not prepared his works for the use of future generations. Under the patronage of Sir Walter Raleigh, England's counterpart of Portugal's Prince Henry the Navigator, Harriot put his mastery of astronomy, cartography, and navigation to use in training the pilots and sea captains employed by Sir Walter Raleigh. In four decades of exceptional scholarship, Harriot had achieved recognition over all Europe and influenced most of his scientific colleagues. But he had seen through the press only one small treatise—the first English account of the new world—*A Briefe and True Report of the New Found Land of Virginia,* 1588. His prodigious output of thousands of pages of mathematical analyses and scientific observations, which had occupied most of his waking hours, were lost to the public, residing in private archives until they were recovered in 1947.

When the career of Thomas Harriot is surveyed, the diversity of his intellect becomes apparent. The present pervasive presence of digital computers makes startingly significant the fact that Harriot first explored the potential of binary numeration. He explored concepts in number theory, combinatorics, linguistic ciphers, and the like, not to mention honing his own sophistication in algebra and spherical trigonometry. Harriot made the earliest telescopic maps of the moon and charted the motions of Jupiter's satellites. He revived the atomic theory of matter (but at a most inauspicious time). And, more to the purpose of our own experiment, made 200 drawings of the sun, carefully charting the movement of spots on its surface in the successful effort to determine the rotation rate of the sun. With such far-reaching perspectives, Harriot stands as a key figure at the time when the new scientific tools of logic, reason, mathematics, and experiment were first being utilized. A man who like his contemporary Francis Bacon took all knowledge for his province, Harriot, through both theory and practice, has proved himself a pivotal character in the emergence of modern science at its birth at the onset of the seventeenth century.

Harriot's Telescopic Observations of Sunspots

Thomas Harriot has left us 200 drawings of the solar surface, all but the first in a numbered sequence. Taking into account the fact that he often observed several times in the course of a day, he has left notes of 450 separate observations. Harriot generally observed soon after sunrise or sometimes just before sunset, although he was prepared to make a drawing for any hour of the day when conditions were favorable.

Harriot conducted most of his observations from the garret of his home at Syon, in the Thames River valley near London. The haze, mists, and vapors near the river in the mornings, as well as the sun's low altitude, provided him a means of diminishing the intensity of the sun's light. For example, his observation of April 19, 1612 has "The sonne being of temperate light in a fayer mist" and on December 13, 1611 there is mention of "thick ayer," a phrase that was subsequently to be often favored and may be descriptive of a slight haze. Clouds were occasionally helpful, but more often were not. "I saw [a cluster of four] twise in a short time through a thinne rag of a cloud, the cloudes quickly obscuring the Sunne." All in all it must be said that Harriot was a patient, diligent, and extraordinarily careful observer.

Thomas Harriot had made several telescopes, which he referred to as "trunckes" or "instruments." In his notes, he differentiates between his various telescopes by designating the magnifying power of each. For example, in the December 3, 1611 sunspot record, Harriot writes,

"Sir William Lower & Christopher saw them with me in several trunckes.
$\frac{10}{1} \cdot \frac{8}{1} \cdot \frac{20}{1}$."

indicating that he used three telescopes with magnifications of 10, 8, and 20, respectively. To steady the telescopes while making observations, Harriot's caption in one of his drawings notes that, "in a payre of stone stayres 2 iron barres beare the Instrumentis."

Sir William Lower had been so impressed with the telescopes and the objects newly seen that he ordered several telescopes from Harriot. In a letter to Harriot in June of 1610, Lower writes:

> We are here so on fire with thes things that I must render my request and your promise to send mee of all sortes of thes Cylinders. My man shal deliver you monie for anie charge requisite. . . . Send me so manie as you thinke needful vnto these obseruations.

In his enthusiasm for making systematic observations of the sun, Harriot was occasionally careless with his sight. In March of 1612 he had looked up at the sun without clouds with his telescope of magnification 20 and then wrote, "My sight was after dim for an houre." The hazards of this sort of work are obvious, and I mention the blindness of Galileo and Cassini by way of warning.

Analysis of the Observations

Out of Harriot's notes, a sequence of drawings has been chosen on which the progress of a distinctive grouping of spots may be followed. These are reproduced in figs. 9.5 a-d. Each drawing is bisected by a vertical reference line. From this line and the time the drawing was made, Harriot also constructed on each drawing a line representing the ecliptic.

Each drawing is accompanied by a record of the time it was made and pertinent remarks regarding visibility or cloud conditions, the magnifying power of the telescope used to make the observations, and the number of spots and their appearances. For our purposes, we will need to pay particular attention to the date and time of the observations. The notation Harriot used in recording his observations is not too complicated; look at the April 19 entry next to Drawing #79.

$$\text{April 19} \quad \odot \quad \text{ho.} \quad 8\tfrac{1}{2} \quad (8\tfrac{3}{4}$$

At the top of the page we see that the drawings were made at Syon in the year 1612, so the date of the drawing, of course, is April 19, 1612. The circle with a dot in its center is the astronomical symbol for the sun; it was a Sunday. (On the April 20 entry you can see a little moon for Monday.) Following the symbol for the day of the week is the abbreviation "ho." for hours followed by the time of day to the nearest 1/4 hour at which the drawing was made. Often Harriot would observe the sun again after he had made a drawing. The time or times he made additional observations are indicated after a left parenthesis. So on April 19, he made Drawing #79 at 8:30 A.M. and made an additional observation later at 8:45. Usually, but not always, Harriott differentiated morning versus afternoon observations by adding P.M. to the time.

The goal of our analysis is to convert the apparent motion of a sunspot along a line in a two-dimensional drawing to the true motion of the sunspot along the arc of a circle on a three-dimensional spherical sun. For example, fig. 9.1 below shows a slice through the sun at the location of a spot.

On the sun, the spot moves from A to B, but seen through a telescope, the spot marches across the circular disk of the sun going from A' to B'. If we can measure how many degrees θ the sunspot swept out in going from A to B (A' to B' on a drawing), then, knowing the number of days D elapsed between A and B, we can determine the sun's period of rotation by

$$P_{syn} = \frac{360° \, D}{\theta}. \tag{1}$$

The subscript 'syn' indicates that this result gives the Sun's synodic period. Harriot was not standing still in space watching the sun rotate. Rather, he was on a moving planet, a planet that was turning in the same direction as the sunspots. We will have to correct for this later.

In order to analyze the motion of a sunspot represented in multiple drawings, we need to transfer its successive images onto a single circle; this will make the sunspot's track across the disk apparent. Choose a *single* sunspot to follow, one that you can identify in at least four or five drawings. On a transparent plastic sheet or a translucent piece of paper, draw a circle with a compass precisely the same size as the sun's disk in Harriot's drawings. If you use transparent plastic, you will need a fine-tipped permanent marker for drawing the circle. Bisect the circle with a dashed line to serve to align the ecliptic in each drawing. This bisected circle will be the template onto which you will transfer the sunspot locations.

Figure 9.1 A slice through the sun at the latitude of a sunspot.

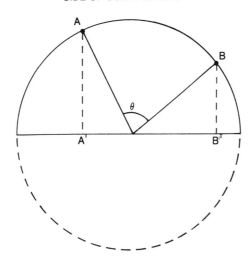

Place your template over the drawing in which the spot you chose is first visible and align the ecliptic on the template to the ecliptic in Harriot's drawing. Place a dot on the template to mark the sunspot's location, and label the location with the corresponding number of Harriot's drawing. Repeat this procedure for each successive drawing until the spot disappears around the opposite limb of the sun. You should find that the spot's track across the disk is along a line. Using a straight edge, draw a line from limb to limb, making the line pass as closely as possible through all the plotted positions. This line represents the projection of the spot's path onto the apparent disk of the sun.

To construct the analog of fig. 9.1 needed for our analysis, measure the length of the line of the sunspot's projection, and on a piece of graph paper draw a line of the same length. Determine the midpoint of the line and construct a semicircle on this line as shown in fig. 9.2. Transfer accurately the sunspot positions onto the straight line segment of the semicircle on the graph paper. An easy way to do this is to overlay the graph paper on the template drawing.

Choose two widely separated sunspot positions and draw a vertical line from each sunspot position to the arc of the semicircle as shown in fig. 9.2. Be careful to avoid using sunspot positions near the limb; a small plotting error by Harriot or by you near the limb can lead to a relatively large error in the determination of the angle θ. The intersection points on the semicircle represent the actual positions of the sunspot on the curved surface of the sun. Connect with a line the intersection points to the midpoint of the straight line segment and measure the angle θ between the two radial lines with a protractor.

Now determine the length of time in days between the observations corresponding to the two sunspot positions you have chosen. Convert each date and time to a decimal date. For example, April 19, 1612, 8 1/2 hours would correspond to the decimal date

$$19 \text{ days} + \frac{8.5}{24} \text{ day} = 19.354 \text{ day}$$

Be careful to add 12 hours (or 0.5 day) if the time was P.M. Record your data in table 9.1.

After you have determined the number of days between the observations, use eq. (1) to obtain the synodic period of the sun.

Repeat your determination of the sun's synodic period for a different spot on the other side of the ecliptic.

Rotation of the Sun

Figure 9.2 Geometrical construction to locate a sunspot on a spherical sun at two different dates. The circle in the bottom portion represents the disk of the sun onto which the location of a single sunspot has been recorded for different dates. The semicircle in the top portion represents a slice through the earth-facing part of the sun at the sunspot's latitude.

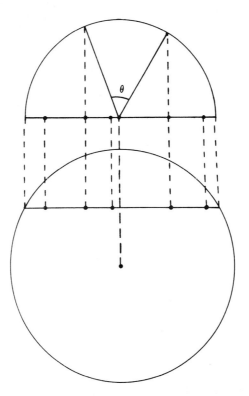

Table 9.1 Data for Determining the Synodic		
	Spot #1	Spot #2
Decimal date of sunspot position #1		
Decimal date of sunspot position #2		
θ		
P_{syn}		

Complications

Because Harriot made his observations from a moving planet, the period required for a spot to make one rotation from earth's perspective is not the same as the period of rotation measured according to some fixed direction in space. In fig. 9.3, when a spot is at S at the center of the sun's disk, the earth is at A. One sidereal period later, the spot turns once and the spot is again at S. But during this time the earth has moved from A to B. From B the spot at S is not yet at the center of the sun's disk. We would have to wait a little longer than one sidereal period for the spot to move to the center of the disk.

To convert the synodic period you obtained to the sidereal period, that is, the period of one solar rotation referred to a fixed direction, use the relation

$$P_{sid} = \frac{P_{syn}E}{P_{syn} + E}$$

Figure 9.3 Geometry for the difference between the sidereal period of a spot and its synodic period.

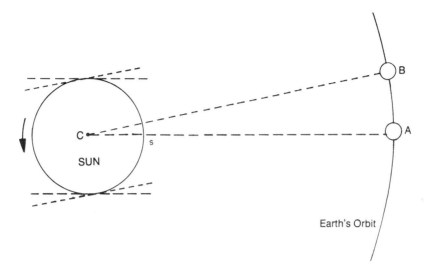

Figure 9.4 Sunspot tracks across the solar disk at various times of the year. PP represents the axis of rotation; N is the north point on the disk.

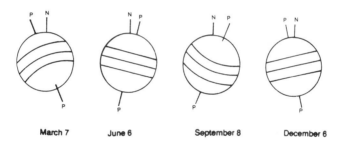

where E is the sidereal period of the earth around the sun, i.e., E = 365.25 days. Calculate the sidereal rotation period of the sun from each of the determinations of the synodic period. Average your results.

$$P_{sid} = \underline{\qquad}$$

P1. The radius of the sun is 7×10^5 km. From your measurements of the rotation period of the sun, calculate the velocity of a point on the sun's equator.

Another complication arises from the fact that the sun's rotation axis is not perpendicular to the plane of the earth's orbit. If the axis were perpendicular to the plane of the earth's orbit, sunspots would always seem to move in straight line paths across the solar disk. Actually this sort of motion is observed only about June 6 or December 6 (see fig 9.4). At other times spots travel in slightly curved paths, the maximum curvature occurring at intermediate dates of March 7 and September 8. As you can see from fig. 9.4, the curvature is not great, and since Harriot's observations used in this lab were made in late April and early May, the sunspot tracks are very nearly straight. No correction need be applied. A careful analysis of sunspot tracks, however, shows that the sun's axis in inclined 7.2° to the perpendicular to earth's orbital plane.

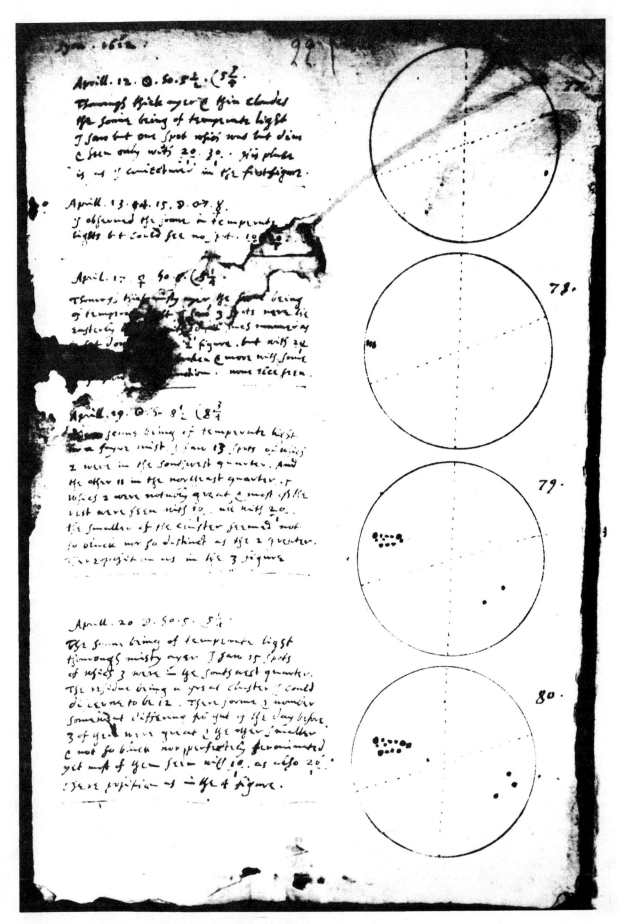

Reproductions from the manuscript copies at the Morris Library University of Delaware and used with the very kind permission of Lord Egremont.

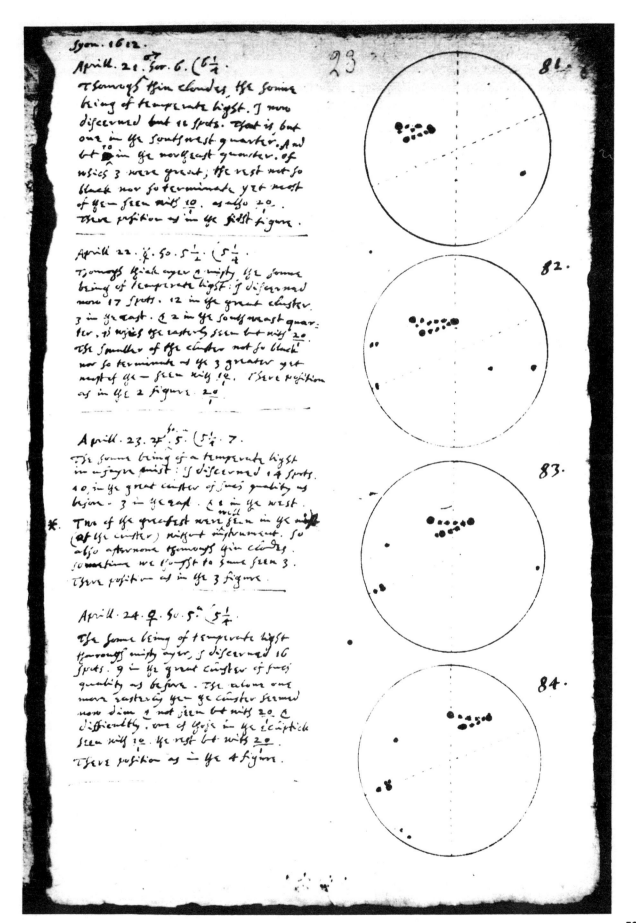

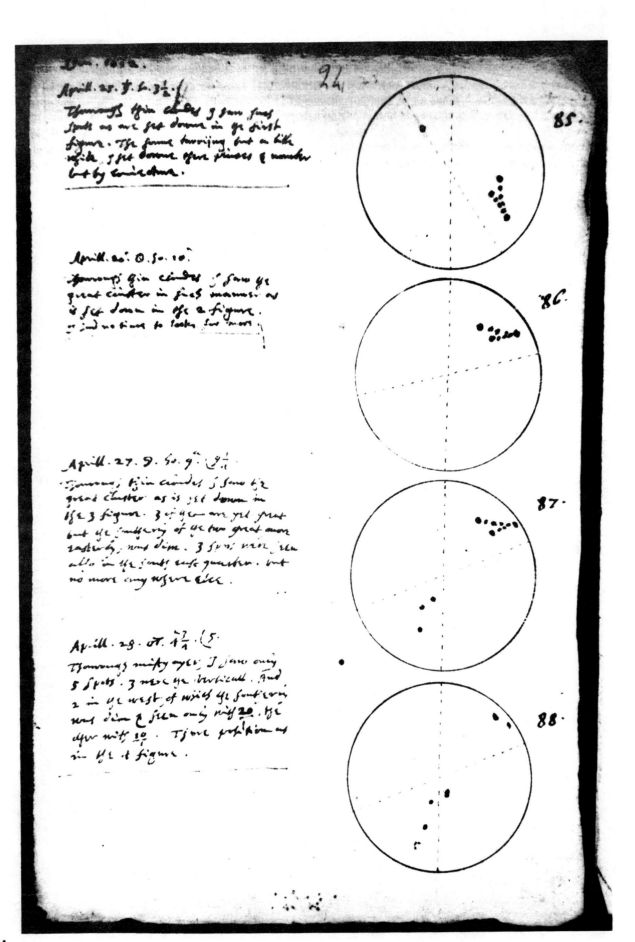

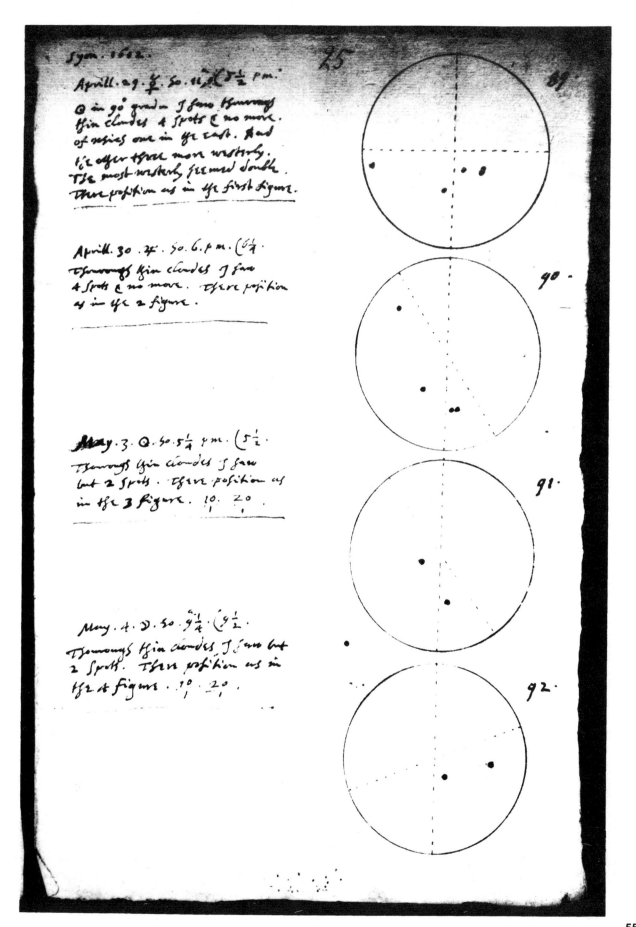

Exercise 10

Radioactive Decay and the Dating of Rocks

Learning Objectives

This experiment measures the half-life of a radioactive element and explores how objects can be dated.

Introduction

From a set of just a few notes, symphonies can be created, symphonies with great complexity, symphonies that are sullen and symphonies that are scintillating.

From a set of just a hundred or so chemical elements, a universe can be made, a universe of stars and a universe of beings.

A particular chemical element is characterized principally by the number of protons in the nucleus of the atom of the element. The number of neutrons in the nucleus, however, may vary from one atom of a chemical element to another atom of the same chemical element, that is, **isotopes** of an element all have the same number of protons but have differing numbers of neutrons. For example, aluminum has only one stable isotope: aluminum–27, or $^{27}_{13}\text{Al}$, with 13 protons and 14 neutrons. Aluminum–26 ($^{26}_{13}\text{Al}$), another isotope, is unstable. It decays radioactively to magnesium–26 by the change of a proton into a neutron with the emission of an anti-electron (e^+, an anti-matter electron having a positive charge) and a neutrino (ν):

$$^{26}_{13}\text{Al} \rightarrow e^+ + \nu + {}^{26}_{12}\text{Mg}$$

The nucleon number (26) stayed the same since the total number of nucleons (protons and neutrons) did not change.

How fast an unstable isotope decays is indicated by its **half-life,** the time it takes for half of the atoms in an initial sample to decay into something else. $^{26}_{13}\text{Al}$ has a half-life of 720,000 years. Suppose you start out with N_o atoms of $^{26}_{13}\text{Al}$. After 720,000 years have passed, only $1/2\ N_o$ atoms would remain; the rest would be $^{26}_{12}\text{Mg}$. Then after another 720,000 years, you would have half of that, or $1/4\ N_o$ atoms. Another half-life later, only $1/8\ N_o$ [$(1/2)^3 N_o$] remains, and so on. The number N of radioactive atoms left after n half-lives is

$$N = (1/2)^n N_o \text{ or } N/N_o = (1/2)^n$$

(n need not be an integer). If information is available regarding the ratio of the number of radioactive atoms present at one time to the number that was present originally, the number of half-lives that have elapsed can be deduced. If the half-life of the isotope can be measured, then the age of the sample can be established. For example, if we know that four half-lives of $^{26}_{13}\text{Al}$ have elapsed since the formation of a certain rock, we would date the rock as being 4(720,000 yrs) = 2.88 million years old.

The chemical elements have widely differing number of isotopes and half-lives. Many heavy elements like polonium (Z = number of protons = 83) and uranium (Z = 92) have no stable isotopes, 13 elements have only one stable isotope, xenon (Z = 54) has nine and tin (Z = 50) has ten. The half-life of beryllium-8 is 2×10^{-16} sec and that of uranium-238 is 4.5 billion years. The half-life of tantalium-180 is estimated to exceed 10 trillion years; that is more than 500 times the age of the universe!

For finding cosmic ages, the more long-lived isotopes are most important. In addition to uranium, elements that can serve as cosmic clocks include rubidium (^{87}Rb), which decays to strontium (^{87}Sr) with a half-life of 47 billion years, and potassium (^{40}K), which decays into the inert gas argon (^{40}Ar) with a half-life of 1.3 billion years. Whatever elements are used, the derived date is the time elapsed since the rocks last solidified.

Let us look at this dating technique in a little more detail. Specifically, from knowing the half-life of a radioactive element, how is the age of the rock determined? From the previous equation, the original number of radioactive nuclei is needed. To get this number, you must know the original composition of the sample rock when it solidified. Often the mineralogical structure of the rock will reveal this. Suppose, for example, a sample contains a mineral that you know is formed with a certain fraction of ^{40}K. The potassium decays to ^{40}Ar. If the argon is trapped in the sample and does not escape, the amount of ^{40}Ar relative to the amount of ^{40}K increases with time. This changing ratio tells the age of the sample.

For example, suppose a rock sample now contains equal numbers of ^{40}K and ^{40}Ar atoms. If there were no argon atoms in the rock originally, they must have all come from the decay of ^{40}K. Exactly half of the ^{40}K nuclei have decayed (and half remain), so the rock must be one half-life old, or 1.3 billion years old. How old is a rock that contains seven times as many ^{40}Ar atoms as ^{40}K atoms? If all the argon came from the decay of potassium, the remaining ^{40}K is 1/8 the original. (Out of eight atoms, seven are ^{40}Ar and one is ^{40}K.) So, three half-lives have elapsed [$1/8 = (1/2)^3$], and the rock must be $3 \times 1.3 = 3.9$ billion years old.

The Experiment

☞ **CAUTION:** *This experiment utilizes a non-regulated amount of radioactive material and is safe when handled properly. In addition to the low level of radioactivity in the solution you will be given, the liquid is also mildy acidic. If the solution comes in contact with your skin, you should wash the area with which the liquid came in contact. Your instructor will tell of some other precautions to take regarding the handling of other equipment used in this experiment.*

In order to illustrate the technique for measuring half-lives, in this laboratory session you will measure the half-life of metastable barium-137. It has a half-life of a few minutes, which is convenient. The metastable barium-137 (denoted ^{137}Ba*) decays into the stable isotope ^{137}Ba by emitting a high energy photon (gamma ray). Though we cannot actually count the number of ^{137}Ba* nuclei present at any time, we can detect the photons that are emitted.

Our eyes can detect only a rather narrow range of photon energies, so a special technique involving a sodium iodide crystal and a photomultiplier tube (PMT) is required to detect and count the high energy photons emitted by the metastable barium-137. In this "scintillation" technique, each photon entering the sodium iodide crystal excites a few thousand electrons in the crystal. When the electrons de-excite, low energy visible photons are produced and seen as a brief flash of light by the PMT placed against the sodium iodide crystal. Thus, the large energy of the gamma ray photon is converted to the smaller energies of many visible light photons. The PMT converts the light flash to a current pulse, which is electronically amplified and counted.

Before beginning the experiment or turning on any equipment, make sure that you have been provided with the proper settings for each device so that the apparatus counts the pulses produced by the gamma ray photons from ^{137}Ba*. The controls on the counter should be set to give automatic 10-second counting intervals, one after another. At some point, either before or after you take readings from your ^{137}Ba* sample, you must time the interval between the beginnings of successive 10-second counts because the next interval does not begin exactly when the preceding one ends. (There is a short period between intervals when the counter resets itself.) That is, ten 10-second intervals will require somewhat more than 100 seconds.

Another measurement you must make either before or after you take readings from the ^{137}Ba* sample is to measure the average number of background counts during a 10-second interval. This average background counting rate must be subtracted from each counting rate of your sample when you analyze your data.

When you receive your sample, place it a few centimeters from the detector and immediately start recording the counting rate in successive 10-second intervals. Continue for about 50–60 intervals.

When finished counting, pour your sample into the waste bottle provided, rinse your beaker in tap water, and wash your hands.

Data Analysis

The rate of detection at a particular time, R(t), follows the exponential decay law

$$R(t) = R_o \exp\left[-\frac{\ln 2}{t^*} t\right],$$

where R_o is the initial detection rate and t^* is the half-life of the radioactive nucleus being studied.

**As an illustration of the effects of differing half-lives, start with 100 radioactive atoms. Using a half-life of 20 seconds, plot the decay rate for times up to 200 seconds after t = 0. You need only plot about 5 points or so to get an idea of what the plot looks like. On the same graph, make another plot of the decay rate of 100 different nuclei that have a half-life of 40 sec, of 100 years, and of 0.02 sec.

The best way to find an accurate value for the half-life of ^{137}Ba* is to plot ln R vs. t. Taking the natural logarithm of both sides of the previous equation gives

$$\ln R = \ln R_o - \frac{\ln 2}{t^*} t.$$

Therefore, a graph of ln R vs. t should be a straight line with slope $-(\ln 2)/t^*$.

Plot this graph, *remembering to subtract the average background counting rate before you compute each natural logarithm.*

Measure the slope of the resulting line and calculate the half-life. Express your answer in both seconds and minutes. Remember that the starting times for successive intervals are a little more than 10 seconds apart.

$t^* = $ _____

P1. A rock known to have no ^{87}Sr when it was formed now contains 1/20 as much ^{87}Sr as ^{87}Rb. How old is the rock?

Exercise 11

Rotation of Planets Using the Doppler Effect

Learning Objectives

In this experiment, you will analyze the spectra of the planets Jupiter and Saturn to estimate their rotational periods.

Introduction

What humans have learned about the universe beyond our planet has come primarily by looking at it. It is true that recently we have embarked on great and heroic adventures to retrieve a few bits of the universe, and, for some time now, we have searched patiently among the rubble of our world for pieces of the universe that have found their way to us. Most of the information about the constituents of the universe, however, has been obtained by the analysis of the electromagnetic radiation they emit.

The light collected from the stars, planets, and other components of the cosmos contains many messages—temperature information, chemical composition, gas pressure, magnetic field strengths, and still other physical parameters. One parameter in particular will be examined in this exercise, namely, the relative velocity of luminous objects. Motion toward or away from an observer is manifested in the source's spectrum. This **Doppler effect,** as it is called, provides a kind of cosmic speedometer that not only tells whether the object is moving toward or away from an observer, but also how *fast* it is moving along the line of sight from the observer to the source.

The Doppler effect is a powerful tool that allows astronomers to find the line of sight velocities of luminous objects without having to know their distance. The Doppler effect is named after the Austrian mathematician, Christian Doppler (1803–1853) who first noted the effect in sound waves. Perhaps you are familiar with this manifestation of the effect. When a source of sound passes by, for example the whistle of an approaching train or the siren on an emergency vehicle, the pitch of the sound changes. It is high when the source is approaching and shifts to a lower pitch as it recedes. Such a shift is a property of *any* wave source in motion and occurs with all kinds of waves—light, sound, even water waves. The French physicist Armand Hippolyte Fizeau (1819–1896) applied Doppler's principle to light waves and recognized the importance of the effect in astronomical applications.

Let us examine in a little more detail the properties of waves and the Doppler effect to see what information about a source's relative motion can be deduced and what cannot.

Waves have three fundamental properties: wavelength, frequency, and velocity. In the water waves pictured in fig. 11.1, the wavelength is the distance between two successive crests of a wave.

> ****How does the distance between two successive crests in the wave compare with the distance between two successive troughs or valleys?

In other kinds of waves, the wavelength is much more difficult to visualize. For light of a particular wavelength, the wavelength is the distance between two successive peaks in electric field intensity. For sound, the wavelength is the distance between two successive peaks in air density. Nevertheless, if you were to plot electric field intensity or air density as a function of distance, you would obtain a curve very much like the curve illustrated in fig. 11.1 for water waves.

The frequency is the number of waves that pass by a point each unit of time, for example, 1/sec (one per sec). The velocity of a wave is the distance covered per unit time by a crest traveling in a certain direction, for example, 1 m/sec.

The three wave properties are related, however, by the relation

$$v = f\lambda \tag{1}$$

where v is the wave velocity, f, the frequency and λ, the wavelength. This fundamental relation applies to all kinds of waves.

Figure 11.1 Snapshot of a moving water wave. The distance between two successive peaks of the wave is the wavelength.

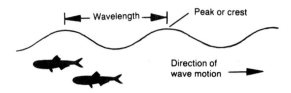

Figure 11.2 Successive crests of light waves emitted by a stationary source S.

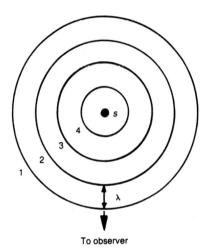

In fig. 11.2, the light source S is stationary with respect to the observer. As successive wave crests are emitted, they spread out evenly in all directions, like the ripples from a stone dropped into a pond. (Let us agree that the circles in fig. 11.2 represent the crests of waves.) The crests approach the observers at a distance λ behind each other, where λ is the wavelength of the light.

Now let us examine what happens when the source of waves moves relative to the observer. When the source is moving with respect to the observer as in fig. 11.3, the successive wave crests are emitted with a source at *different positions*, S_1, S_2, S_3, and S_4. Thus, to observer A, the waves seem to follow each other by a distance *less* than λ, whereas to observer C they follow each other by a distance *greater* than λ. The result of the motion of the source is that the wavelength of the radiation received by A is shortened; the radiation received by C is lengthened. To a distant observer B at right angles to the motion of the source, no effect is observed. The effect is produced only by motion toward or away from the observer, a motion called **radial velocity**. Observers between A and B, and between B and C would observe some shortening or lengthening of the light waves, respectively, for a component of the motion of the source in their line of sight (see fig. 11.4).

It is easy to see from fig. 11.3 that the faster the source is moving, the more the wavefronts get piled up ahead of it and the more the wavefronts behind the source get drawn apart. The relation between the relative radial velocity v of the source and the shift in wavelength Δλ is

$$\frac{v}{c} = \frac{\Delta\lambda}{\lambda_0} \qquad (2)$$

where λ_0 is the wavelength of the light if the source were stationary, and c is the speed of light (c = 3 × 10^8 m/sec).

Figure 11.3 The Doppler effect produced by a moving source of waves.

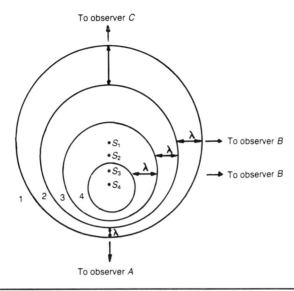

Figure 11.4 Only the component of velocity along the line of sight is measured by the Doppler effect.

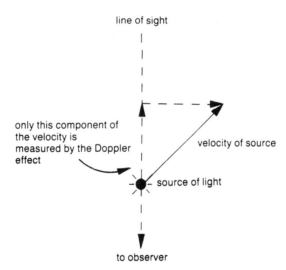

The Experiment

Part 1 The Rotation of Jupiter

The Doppler effect, as you can see, measures velocity along the line of sight. As a prelude to measuring the rotation period of the planet Jupiter, the length of its "day," let us think about what we should expect if we measured the equatorial velocity of the planet.

Figure 11.5 is a view from above a rotating planet. As the planet rotates, a portion of the planet's limb (its edge) is moving toward the observer; the opposite side of the planet's disk is moving away from the observer. At the center of the planet's disk, the motion is across the line of sight. In terms of the Doppler effect, the wavelengths of the light from the approaching limb will all be shortened, while the wavelengths of the light from the receding limb will all be lengthened. The wavelengths of the light from the center of the disk, which is moving across the line of sight, will be unaffected.

Figure 11.5 Relative velocities of a rotating planet's equator.

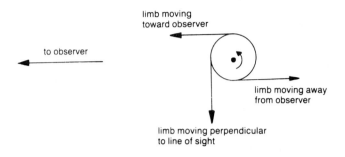

Figure 11.6 Doppler effect on an absorption line of wavelength λ_0 in the spectrum of a rotating planet.

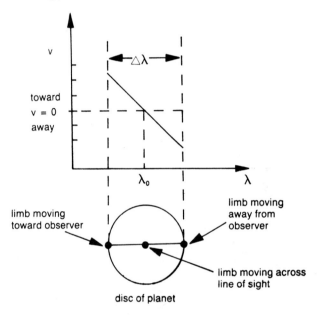

You will be analyzing the spectrum of light from Jupiter and will be measuring the Doppler effect on absorption lines in the spectrum. In view of the above, we might expect something like that pictured in fig. 11.6.

Our ultimate purpose will be to determine the period of rotation of Jupiter, but a few comments are necessary regarding the relation of the equatorial velocity of the planet to the relative velocity of the limbs as observed from the spectra; the two are *not* identical. The sun, earth, and planet are all moving relative to each other. The relative velocity of the limbs depends not only on the equatorial velocity, but also on the angle at the planet between the earth and sun. But when Jupiter is at opposition (see fig. 11.7), the relative *radial* velocity of all the three bodies is zero.

In the case of a planet shining by reflected light, the displacement of the spectral lines is the sum of the geometrical rotation effects both with reference to the sun, as the source of light, and with reference to the observer. When the planet is at opposition, as in our case, the displacement is just doubled by the reflection. In addition, since one limb of the planet is receding and the other approaching, the total displacement from one end of a spectral line to the other represents twice the equatorial velocity. *The result of these two effects is that the measured value of $\Delta\lambda$ (see fig. 11.6) gives a value for v that is four times the equatorial velocity of the planet.*

Analysis of the Spectrum

☞ **Note:** *In order to make precise measurements on Jupiter's spectrum, you will need a sharp pencil or something similar that can draw thin sharp lines. You will also need a see-through ruler that is graduated in millimeter divisions (or smaller).*

Figure 11.7 Jupiter at opposition. At opposition Jupiter is neither moving toward nor away from the sun and earth. Earth, also, is neither moving toward nor away from the sun and Jupiter.

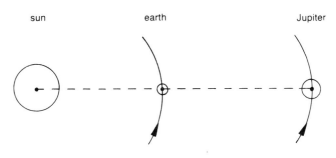

Figure 11.8 Spectrum of Jupiter with iron comparison lines.

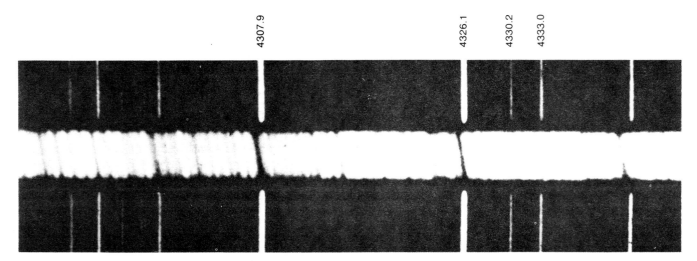

A portion of Jupiter's spectrum appears in fig. 11.8 as the central broad band crossed by dark absorption lines. On either side of Jupiter's spectrum are a few bright comparison emission lines due to iron; the iron lines provide a reference for measuring wavelengths on fig. 11.8.

As the first step in our goal to measure the equatorial velocity of Jupiter, we need to measure the difference in wavelength of a single absorption line from one limb to the other. We begin this by accurately delineating the limbs. Place a clear straight edge precisely along the top of the broad band representing Jupiter's spectrum, being careful not to be distracted by the presence of the absorption lines, and draw a thin sharp line to define the limb of the planet. Do likewise for the bottom of the broad band representing the opposite limb of Jupiter.

Pick an absorption line in Jupiter's spectrum and with a clear straight edge draw a thin sharp line precisely down the middle of the absorption line. This line down the middle of a chosen absorption line intersects the lines representing the limbs of the planet in two points.

To determine the two wavelengths corresponding to the two intersection points, we will need to use the comparison spectrum. The wavelengths of the comparison lines are indicated in angstroms on fig. 11.8 (1 Å = 10^{-8} cm, but you will not need to convert). Pick a labeled comparison line relatively *far away* from the absorption line you have chosen and measure the distance to each of the intersection points on the limb from the comparison line. Some comparison lines are rather broad, so measure from the centers of the comparison lines. Try to estimate the distances to tenths of a millimeter. Subtract the smaller distance from the larger; call the difference Δx.

To convert the difference in the two distances to a difference in wavelength, we need to establish how distance and wavelength are correlated on the spectrum. Again we will use the comparison lines. Measure the distance between two

Table 11.1 Estimates of Jupiter's Equatorial Velocity

Absorption Line	Δλ	λ	$v_{equatorial}$
1			
2			
3			
4			

widely spaced, labeled comparison lines and find the wavelength difference between the two comparison lines you have chosen. The dispersion of the spectrum is

$$\text{dispersion} = D = \frac{\text{wavelength difference between comparison lines}}{\text{distance between comparison lines}} \qquad (3)$$

The wavelength difference between an absorption line at the receding limb and the line at the approaching limb is

$$\Delta\lambda = D\Delta x \qquad (4)$$

Enter your result in table 11.1.

Estimate the wavelength λ of the absorption line you have chosen. You only need a precision of 5 or 10 Å so you can make the estimation by eye. Enter your result in table 11.1.

Calculate the velocity corresponding to the Δλ and λ you have measured by means of the Doppler effect relation

$$v = \frac{\Delta\lambda}{\lambda} c, \qquad (5)$$

where c = speed of light = 3×10^8 m/sec. Divide the v so obtained by 4 to obtain your estimate of the equatorial velocity. Enter this result in table 11.1.

Choose three other absorption lines in Jupiter's spectrum and repeat the previous procedure to obtain three additional estimates of Jupiter's equatorial velocity. Record your measurements in table 11.1.

P1. Determine the average value of your estimates of Jupiter's equatorial velocity.

$$v_{equat.} = \underline{\hspace{3cm}}$$

For a point moving with constant speed, the distance traveled by the point is the speed multiplied by the time the point is moving. To get Jupiter's rotation period, therefore, we need to calculate

$$\text{period} = \frac{\text{circumference of planet}}{\text{equatorial velocity of planet}}$$

or

$$P = \frac{2\pi r}{v_{equat.}} \qquad (6)$$

The equatorial radius of Jupiter is 7.085×10^7 m.

P2. Calculate Jupiter's period of rotation using eq. (6) and your result from P1. Express your answer in hours.

$$P = \underline{\hspace{3cm}}$$

Figure 11.9 Spectrum of Saturn and its rings with neon comparison lines. (Courtesy H. Spinrad)

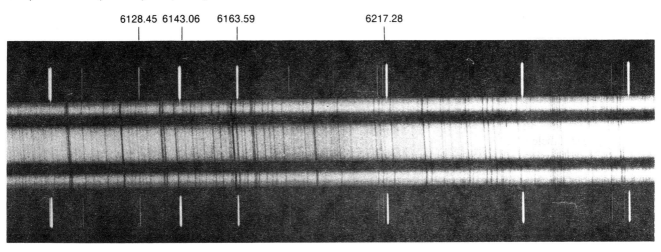

Part 2 The Rotation of Saturn

Figure 11.9 displays the spectrum of Saturn and its rings. The central wide band is the spectrum from the planet, and the two narrower bands on either side are from the ring system. As in the case with Jupiter, Saturn's spectrum was recorded when the planet was at opposition. Additionally, the ring system was very nearly along our line of sight and appears almost edge-on. A neon comparison spectrum appears in fig. 11.9 to establish a wavelength scale.

The spectrum of the ring system is certainly different from that of the planet. As we saw previously, a planet rotates as a solid body giving rise to tilted absorption lines. The rings of Saturn are not solid, but are composed of small particles—a myriad of moons. Just as in the case of the solar system where an inner planet revolves around the sun faster than does an outer planet, the inner ring particles revolve faster than outer ring particles. This is opposite of how a solid body rotates, so the ring absorption lines tilt opposite to the planet's. You will be working exclusively with absorption lines from the planet, and not the rings.

Using the procedure that you used for Jupiter, obtain a value for Saturn's rotation period. Use table 11.2 to record your measurements from Saturn's spectrum.

P3. Determine the average value of your estimates of Saturn's equatorial velocity.

$$V_{equat.} = \underline{\hspace{2cm}}$$

P4. Saturn's equatorial radius is 6.003×10^7 m. Calculate Saturn's period of rotation, and express your answer in hours.

$$P = \underline{\hspace{2cm}}$$

Table 11.2 Estimates of Saturn's Equatorial Velocity

Absorption Line	Δλ	λ	$V_{equatorial}$
1			
2			
3			
4			

P5. Calculate the equatorial velocity of the earth and compare it to the equatorial velocities of Jupiter and Saturn (the equatorial radius of the earth is 6,378 km).

Exercise 12

Power Output of the Sun

Learning Objectives

This experiment measures the rate at which the Sun is emitting energy.

Introduction

The sun powers the earth and the life that flourishes upon it. It warms us and feeds us, and illuminates our world. Life on earth is intimately tethered to the energy radiated by the sun. Birds greet the sunrise with audible ecstasy, and even some one-celled organisms know to swim to the light. Think of the sun's heat on your upturned face on a cloudless summer's day; think how dangerous it is to gaze at the sun directly. From 150 million kilometers away, we recognize its power. It is powerful beyond human experience. Our ancestors worshiped the sun, and they were far from foolish.

For all its furious energy output, the sun is a rather reliable source of energy. Some of the best measurements of the rate of energy emission from the sun were made by instruments aboard the spacecraft called the Solar Maximum Mission. From above the earth's atmosphere, it was able to measure the sun's rate of energy emission with high precision, and it found variations of about 0.1 percent over the course of a few weeks. These variations appear to be related to changing numbers of sunspots. Such short-term variations do not seem to have much of an effect on the earth or its weather, but a 0.1 percent variation that lasted for a decade might have a very significant impact, especially on world agriculture. A variation of only 1 percent would change the average temperature of the earth by 1–2°C (about 1.8–3.6°F). For comparison, during the last ice age, the average temperature on the earth was about 5°C cooler than it is now.

What could possibly be the source of the sun's energy that could have sustained its prodigious energy output for so long? During the late nineteenth century, when the earth was found to be billions of years old, the source of the sun's power became an embarrassingly difficult puzzle to explain. The crux of the problem was how to explain both the rate of energy production and its longevity. The sun has been roughly the same luminosity for at least 3 billion years. How do we know? Geologists have found rocks containing fossils of living organisms that are at least that old. For life to exist, the earth must be warm, and, therefore, the sun must have been roughly at its present luminosity.

What about chemical reactions as the source of the sun's energy? Energy from ordinary chemical reactions, such as burning, could not explain both the energy production rate and the time scale over which that rate must be maintained. If the sun were composed entirely of oxygen and coal (carbon), to maintain the observed energy output, the one you will measure, the sun would have burned to a dark cinder in about 20,000 years. Obviously, the sun could not be a coal-burning furnace!

In the middle of the nineteenth century Herman von Helmholtz (1821–1890) and William Thomson Lord Kelvin (1842–1907) proposed that the sun shone because it was releasing gravitational energy by shrinking, that is, gravitational contraction converted gravitational potential energy to radiative energy. As the sun contracts, the gases get squeezed, with the result that the temperature rises with a concomitant emission of radiation.

Because of the sun's substantial mass, a contraction rate of only 40 m/year would liberate the required energy. The gravitational energy stored in the sun would last for about 20 million years, far longer than the age of the earth as determined by early geologists.

P1. The sun has a diameter of 1.4×10^9 m and its distance is 1.5×10^{11} m.
   a) Verify that the *angular* diameter of the sun is $0.53°$.
   b) Assuming that gravitational contraction as described above is responsible for the energy output of the sun, by how much would angular diameter of the sun have decreased over a thousand years? Typical accuracies for an angular measurement in ancient times ($\sim 2{,}000$ years ago) was a few minutes of arc (~ 5). Could we have noticed a decrease in the size of the sun from an ancient measurement?

Near the beginning of the twentieth century, the study of nuclear transformations indicated that the energies involved in the atomic nucleus were a million times larger than the energies involved in chemical reactions, which were due to the rearrangement of the outer electrons among atoms. Here was the possibility of a tremendous energy source.

Albert Einstein provided the key idea about the sun's energy. Grappling with the fundamental nature of electromagnetic waves, he demonstrated that mass and energy are related by the equation $E = mc^2$, where E is the energy (in joules) released in the conversion of a mass m (in kilograms) and c is the speed of light (in meters per second).

But how to change matter into energy?

Not until the 1930s did the mechanism by which the sun generated its energy from nuclear reaction become understood. The energy comes from the same reactions that power a hydrogen bomb. They are called **fusion reactions** because they fuse nuclei together.

A number of different fusion reactions generate the sun's energy, but one series of reactions having the net effect of fusing four hydrogen nuclei to make a single helium nucleus accounts for more than 85 percent of the energy produced by the sun. One helium nucleus has 0.7 percent less mass than four hydrogen nuclei, and most of the mass difference appears as energy.

4 hydrogen nuclei	..........6.693×10^{-27} kg
1 helium nucleus	..........6.645×10^{-27} kg
difference in mass	..........0.048×10^{-27} kg

Using Einstein's equation

$$E = mc^2$$
$$= (0.048 \times 10^{-27} \text{ kg})(3 \times 10^8 \text{ m/s})^2$$
$$= 4.3 \times 10^{-12} \text{ Joules}$$

In this experiment you will measure the power output of the sun. Your answer will be expressed in units of watts (energy/time) or, equivalently, in Joules/sec. Once you know the power output of the sun in Joules/sec, you can determine the number of reactions per second needed to maintain the observed power level. Finally, you will be able to estimate the lifetime of the sun from your observations.

The Experiment

Imagine a source of light, say a 100-W bulb. The 100 W means that the bulb emits 100 Joules of energy each second (1 W = 1 J/sec). Now imagine this bulb at the center of a sphere 2 m in radius. The energy emitted by the bulb in the form of electromagnetic radiation is spread out over the surface of the sphere. The intensity of the light, the power per unit area, is

$$I = \frac{P}{A} = \frac{100 \text{ W}}{4\pi(2 \text{ m})^2} = 2 \text{ W/m}^2$$

If we imagined another sphere surrounding the light bulb with a radius of 4 m (twice 2 m), the intensity of the light on this surface would be

$$I = \frac{P}{A} = \frac{100 \text{ W}}{4\pi(4 \text{ m})^2} = \frac{1}{2} \text{ W/m}^2$$

one-fourth what it was at 2 m. We can generalize these examples by noting that the intensity of light falls off as the square of the distance from the source.

In this experiment, you will use a null photometer to balance the power levels of sunlight and the light from a standard light bulb.

The null photometer is a box open at both ends that allows the light intensity from two different sources to be compared and balanced.

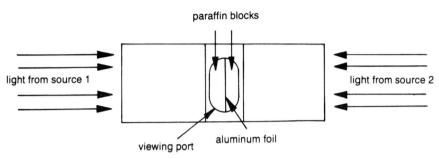

At the center of the null photometer are two blocks of paraffin cemented to sheets of aluminum foil. The light from some source enters the photometer and gets scattered inside the paraffin. Some of the light gets scattered to the side of the paraffin block where it can be viewed. The aluminum foil prevents light from one source from leaking over to the other side of the photometer where it would interfere with the light level measurement of the second source. The aluminum foil also reflects light from the source, giving a greater efficiency to the scattering process.

In making a measurement with the null photometer, the relative positions along a line of the two sources and photometer are adjusted so that the scattered light intensity from both paraffin blocks are equal as seen from the viewing port.

The human eye/brain detection system is quite good at matching light levels and perceiving small differences in light intensity. When color differences are present between the light from the two sources, the balance is more difficult to achieve and reproduce. To remove to a degree the effects of color differences, a piece of colored plastic is placed over the viewing part so that the eye is balancing the same color.

Part 1

To verify the experimental technique we will use to measure the power output of the sun, we will first perform a controlled experiment in the lab. Arrange two light bulbs of differing power outputs along a line, as shown below, with the null photometer placed between them.

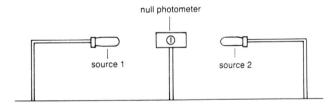

Be careful to place all the components at the same height and along a line.

Adjust the relative positions of the sources and photometer so that the same intensity is registered on both halves of the photometer's viewing port. Measure the relative distances of source 1 to the photometer and from source 2 to the photometer. Measure from the center of the bulb's spherical portion to the center of the null photometer.

When the null photometer is balanced, the intensity of light of source 1 is the same as the intensity of source 2.

$$\frac{P_1}{4\pi(d_1)^2} = \frac{P_2}{4\pi(d_2)^2}$$

P2. Using one source as a standard, and the relative distances you measured, predict what the power output of the second source should be. Check to see whether the rating on the second bulb corresponds to your prediction.

Part 2

The intensity of sunlight is not easy to match with a light bulb, but the null photometer can be modified so that it is able to compare two different light sources with very large differences in power output.

The intensity of a very bright light source can be reduced by stopping down the end of the photometer pointing to the bright source. The reduction can be achieved by diminishing the size of the opening from which light can reach the paraffin.

Set up the two sources and photometer as in Part 1 of this experiment, except on the side of the photometer toward the brighter source attach the mask that lowers the effective area of the null photometer when compared to the area A_2 of the other side of the photometer. The side toward the brighter source will be measuring a power output diminished by a factor of A_1/A_2, where A_1 is the area over which light is admitted into the stopped-down portion of the photometer.

[A diffuser is mounted over the hole in the mask. By diffusing the light reaching the paraffin, a more uniform illumination is achieved.]

P3. Verify that the modified null photometer is still able to predict the value of the power output of the fainter source.

Part 3

Take the null photometer with the aperture stop and the brighter light bulb outside and measure the power output of the sun. (The distance to the sun is 1.5×10^{11} m.)

Power output of the sun = _____

P4. From the introduction, it was seen that one fusion reaction produces 4.3×10^{-12} J. How many reactions each second are required to sustain the power output of the sun that you measured?

P5. How much mass is the sun losing each second in making this mass-to-energy conversion?

P6. The sun's mass is 2×10^{30} kg and is 75% by mass hydrogen. Because the nuclear reactions occur only near the hot, dense core, only about 15% of the total amount of hydrogen is available for fusion reactions. Estimate the total energy available for radiation. (There are 6×10^{26} hydrogen nuclei in 1 kg of hydrogen.)

P7. From the total energy available, estimate the lifetime of the sun assuming the sun has been radiating the power level you measured over its entire lifetime.

Exercise 13 Photometry

Learning Objectives

In this experiment, you will analyze the brightness of objects in different regions of the visible spectrum and deduce the relative temperatures of the objects from the brightness measurements.

Introduction

Electromagnetic radiation—light—is a by-product of many different atomic processes. The radiation that is produced is characteristic of the process producing the radiation. Here is the handle that astronomers have on the stars. Even though removed by imponderable distances, astronomers can deduce some important physical characteristics of the systems they study by examining the radiation emitted from those systems. Sometimes the radiation may be difficult to detect—being too feeble or in an exotic region of the spectrum—but humans have been rather clever and have overcome these difficulties to a large degree. The result is that nearly the entire electromagnetic spectrum has been opened to our study. Now we can see the signals of the stars.

You are probably aware of one phenomenon that produces radiation—hot objects glow. Place a nail in a gas flame and soon it glows deep red, and it is not very bright. Keep it in the flame a little longer and the color changes to an orange hue. The nail also is noticeably brighter than before. Keep it in the flame long enough and it will eventually become a bright white (if the flame is hot enough). Even before the nail glows visibly, when the nail has been in the flame only a short while it is emitting radiation. If you put your finger near the heated nail, you can sense the energy, the heat being radiated by the nail. This is similar to placing your hand near an iron; the invisible radiation that is emitted is infrared light, and it so happens that human eyes do not detect this radiation.

It is clear from observations of hot objects that both the color and amount of light emitted from an object depends on the object's *temperature*.

Experimental Diversion No. 1 (Optional)

The visible region of the spectrum is often quoted as running from 360 nm (1 nm = 10^{-9} m) to 720 nm. This range represents a sort of average among many people. As you might guess, different people have slightly different ranges of sensitivity to light.

A monochromator can produce and deliver light of nearly a single wavelength and vary the wavelength of the light that is delivered. If a monochromator is available, check the wavelength extremes for your range of vision. Compare your range with others. Is there more variation on the blue end of the spectrum or the red end?

Sometime during the lab session, when the monochromator is available, you and your partner might try the following:

1. Find the wavelength of pure "blue." Compare your result with your partner's.
2. Try the same for pure "green."
3. Try it for pure "yellow."
4. Try it for pure "orange."

What do you think the results mean?

Experimental Diversion No. 2

You should have available for your use a 100-W bulb made of clear glass and a Variac. The Variac can be adjusted to deliver any voltage between 0 and 120 V. The power ratings on light bulbs (the "100 W" in our case) are applicable when operated at 120 V, the usual voltage (nominally) provided by standard electrical outlets. If the light bulb is operated at a voltage lower than the standard 120 V, the light bulb's power output is smaller. (The power output is approximately proportional to the square of the voltage at which it is operated.) The Variac is a convenient means to change the voltage applied to a light bulb.

Connect the light bulb to the output of the Variac. Beginning at 0 V, increase the voltage slowly and note qualitatively the amount of heat, the amount of light, and the color of light produced.

Be careful not to actually touch the glass dome of the bulb, especially when operated at the higher voltages. The glass gets unpleasantly hot.

At all temperatures, objects are emitting and absorbing radiation. If an object's temperature is neither rising nor falling, then the amount of radiant energy absorbed per second is the same as the amount of energy radiated per second. For real objects details about the object, such as surface texture and color, influence these radiative processes. It is possible to imagine an ideal object, called a **blackbody,** whose radiative properties do not depend on such details. From a theoretical standpoint, the physical properties of a blackbody can be deduced from thermodynamics, but beyond this, as it turns out, many objects of interest behave very much like a blackbody—stars, for instance.

Two related questions we want to answer are:

1. How does the amount of energy radiated vary with temperature?
2. For an object at a fixed temperature, how does the brightness of the object differ in different wavelength regions?

The third quarter of the last century witnessed extended efforts to answer such questions. It was found empirically, that is, from brute force measurements not based on fundamental theoretical guiding principles, that the amount of power (energy per time) radiated by a blackbody depended only on its temperature. (Laboratory systems can be constructed that behave very nearly like blackbodies.) In fact, the power radiated per unit area of the blackbody E depended on the 4th power of the temperature

$$E = \sigma T^4 \tag{1}$$

where σ is a constant; $\sigma = 5.67 \times 10^{-8}$ W/m²K⁴. The temperature T is measured on the fundamental **kelvin scale.** Double the temperature of an object, and you increase the amount of energy radiated by a factor of 16!

After the development of the quantum theory in the first quarter of this century, the **fundamental physics underlying** eq. (1) were understood. In fact, the empirical constant σ in eq. (1) was shown to have structure, that is, it is a **combination** of fundamental constants of nature (like the speed of light, Planck's constant, etc.).

In the experiments near the end of the last century, which examined the radiant energy emitted by hot objects, it was discovered that for an object at a specific temperature, the amount of energy radiated, the brightness of the object, depended on the wavelength at which it was examined. The distribution of energy radiated as a function of wavelength was characterized by a peak at a well-defined wavelength. (See fig. 13.1.) Furthermore, the wavelength at which the peak λ_{max} occurs varies with the temperature of the object: the hotter the object, the smaller λ_{max} is. This behavior was quantified near the end of the last century in the empirical relation where T is measured on the absolute (kelvin) scale. This behavior is illustrated in the spectral distribution curves in fig. 13.1. Notice that:

$$\lambda_{max} = \frac{2.9 \times 10^{-3} \text{ m-K}}{T} \tag{2}$$

1. For increasing temperatures of objects, the peak in the curve moves toward shorter and shorter wavelengths, i.e., the wavelength at which an object is brightest moves to smaller values.
2. An object with a higher temperature emits more energy *at each wavelength* than an object at a cooler temperature. Therefore, the hotter object is brighter than the cooler one.
3. Some energy is emitted at all wavelengths forming a smooth, continuous spectrum.

Stars, while not precisely blackbodies, do come quite close. Judge for yourself. Figure 13.2 gives the spectral distribution of energy as a function of wavelength for the sun (solid line). The data represented are satellite data obtained outside the atmosphere and, consequently, unaffected by any absorptions and distortions produced by the gases in our atmosphere. The sun's spectrum, of course, has its own absorption lines produced by atoms in its own atmosphere. The atoms absorb radiation at specific wavelengths, thus producing the absorption lines. The energy is re-radiated at *different* wavelengths, distorting the smooth blackbody spectrum. Nevertheless, you can see from fig. 13.2 that the sun's spectrum still is very similar to that of a blackbody.

Figure 13.1 Energy emitted as a function of wavelength for laboratory blackbodies at various temperatures.

1μm = 10⁻⁶ meter.

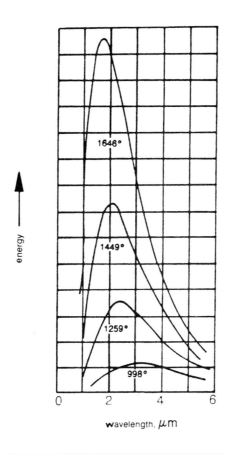

Figure 13.2 Energy emitted as a function of wavelength for the sun (solid line). A blackbody curve (dashed line) for T = 6000 K is shown for comparison. 1μm = 10⁻⁶ meter.

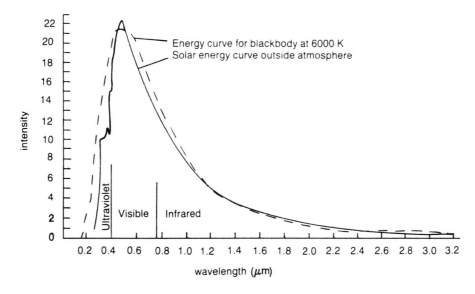

Photometry 73

Figure 13.3 Schematic blackbody curves for stars at different temperatures.

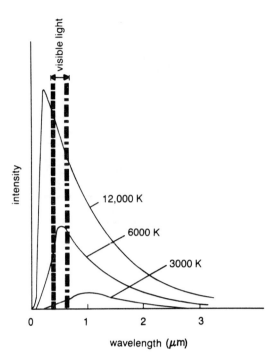

P1. From fig. 13.2, determine the wavelength of the peak in intensity for the sun (solid line). Use eq. (2) to determine what blackbody temperature the wavelength corresponds to.

The spectral curves for stars at three different temperatures are shown in fig. 13.3. The visible region of the spectrum is marked; the blue region is to the left, and the red is to the right. Look at the spectral curve for the 12,000 K star.

The spectral curve of the 12,000 K star is much higher in the blue than in the red. That is, if I look at such a star through a blue filter and a red filter, the star would be brighter through the blue filter than through the red.

The situation is reversed for the 3,000 K star. The red intensity is greater than the blue intensity. This star appears red.

Let us quantify this somewhat by looking at the *ratio* of the intensities in the blue and red for different temperatures and compare them. From what we have already said,

$$\frac{I_{blue,12,000}}{I_{red,12,000}} > \frac{I_{blue,3,000}}{I_{red,3,000}}$$

You can see that the value of I_b/I_r is a measure of the temperature of a star; the larger the ratio, the hotter the star. Let us also refer to the ratio as the *color* of the star.

The Experiment

In order to quantify the spectral characteristics of a source of light, you will use a photodetection system comprised of a photocell and a set of filters. The photocell operates when light strikes its photosensitive surface. As a result, the photocell produces a measureable electrical current proportional to the intensity of light striking the cell: double the intensity and the current doubles. This is how solar cells operate. An ammeter in the circuit measures the current produced by the cell. Switches on the meter change the sensitivity of the meter so that a wide range of light intensities can be measured.

Table 13.1 Data for Color vs. Distance Analysis

Distance (cm)	Filter	Source & Background (Milliamps)	Background (Milliamps)	Source (Milliamps)
30	blue			$I_b =$
	yellow			$I_y =$
60	blue			$I_b =$
	yellow			$I_y =$
90	blue			$I_b =$
	yellow			$I_y =$

Part 1 Does the Color of a Source Depend on Distance?

For this portion of the experiment, we will examine a source using blue and yellow filters. You can probably guess the answer to the question posed above, but let us check to see how I_b/I_y behaves with distance from a source.

Measure the intensity of the 100-W source through blue (b) and yellow (y) filter for source-detector separations of 30 cm, 60 cm, and 90 cm. To do this, place the photocell detector at the appropriate distance from the bulb and cover the photocell with a filter. Now turn off the light source or cover it with opaque black paper. The small current still being registered by the meter is due to electronic noise in the circuit and stray light in the room. Record this *background* level in table 13.1. Turn on the light or uncover it. The current now registered by the meter corresponds to *source and background*. Enter this reading in table 13.1. To obtain the intensity of the source alone, subtract the background reading from the source and background measurement. Repeat this procedure for each filter and at each distance indicated.

Calculate the ratio I_b/I_y for each detector distance.

    $d = 30$ cm $I_b/I_y = $ _____
    $d = 60$ cm $I_b/I_y = $ _____
    $d = 90$ cm $I_b/I_y = $ _____

P2. Qualitatively, how does I_b vary with distance? I_y?

P3. How does the ratio I_b/I_y behave with distance?

Part 2 Temperature-Color Correlation

To investigate how the color ratio I_b/I_y changes with temperature, we will examine the light emitted by the filament of a 100-W light bulb as the voltage supplied to the bulb is varied. You can adjust the voltage supplied to the bulb, and, consequently, the power output of the bulb by connecting the bulb to a Variac. So, by changing the voltage at which the bulb is operated, we can change the filament temperature.

Connect the bulb to the Variac as described and adjust the Variac to supply 30 V. The filament should glow with a noticeable orange color. Place your photo detector at a distance of 15–20 cm away from the filament; the source should be bright enough so that a measureable current is produced above any background level. You may have to adjust the scale on the meter to achieve the maximum number of digits for the meter's display.

Keeping a fixed distance, proceed as in part 1 measuring the intensity of the background and the source plus background for each filter and each voltage beginning at 30 V and increasing the increment by 20 V up to 110 V. Record your data in table 13.2.

Table 13.2 Data for Temperature-Color Correlation

Voltage	Filter	Source + Background	Background	Source	Ratio I_b/I_y
30	b			$I_b =$	$I_b/I_y =$
	y			$I_y =$	
50	b			$I_b =$	$I_b/I_y =$
	y			$I_y =$	
70	b			$I_b =$	$I_b/I_y =$
	y			$I_y =$	
90	b			$I_b =$	$I_b/I_y =$
	y			$I_y =$	
110	b			$I_b =$	$I_b/I_y =$
	y			$I_y =$	

P4. Qualitatively, how does the color ratio I_b/I_y vary with temperature?

P5. Qualitatively, how does the brightness of the filament vary with temperature?

Part 3 Color of the Sun (Optional)

☞ *Be careful never to look directly at the sun.*

If the weather is clear, you can measure the color ratio I_b/I_y for the sun. You can do this indoors if you have a clear view of the sun or outside, if practical.

In making the background measurements as before, it is not easily possible to cover up the source, so place a large opaque object a meter or so away from the detector and allow the shadow of the object to fall on the detector.

Determine the color ratio I_b/I_y for the sun.

P6. Is the sun hotter or cooler than the lamp filament operated at 110 V? How do you know?

Part 4 Comparison of Radiative Output of Sunspots vs. Photosphere

Table 13.3 gives the intensity per unit area at various wavelengths for sunspots on the sun and for unspotted regions of the sun's visible surface (the photosphere). It is possible to construct filters that transmit only a narrow region of the spectrum, a region fractions of a nanometer in width. The blue filter you used previously may have transmitted significant amounts of light in the wavelength range 380 nm–420 nm. The filters used to obtain the data in table 13.3 permitted transmission in a much narrower band, typically 1/2 nm on either side of the specified wavelength.

Table 13.3 Radiation Intensity per Unit Area from Photosphere and Sunspot

Wavelength (nm)	Intensity Sun	Intensity Sunspot	Wavelength (nm)	Intensity Sun	Intensity Sunspot
300	10.1	2.1	550	106.1	37.1
325	23.9	5.3	600	95.1	35.2
350	50.6	11.6	700	68.7	28.1
375	65.3	16.3	800	50.2	23.0
420	111.6	31.2	1000	31.3	17.2
450	117.0	35.1	1100	25.1	14.8
470	117.7	36.5	1200	20.7	12.6
500	114.4	36.6			

From the data in table 13.3, determine the approximate wavelength of maximum intensity for both the sunspot and the photosphere. Using eq. (2), calculate the temperatures corresponding to those peaks in intensity.

λ_{max} (photosphere) = _____   T (photosphere) = _____

λ_{max} (sunspot)   = _____   T (sunspot)   = _____

Use eq. (1) to determine how many times more energy per unit area is being radiated by the photosphere than by a sunspot, i.e., calculate

$$\frac{E_{photosphere}}{E_{sunspot}} = \frac{\sigma(T_{photosphere})^4}{\sigma(T_{sunspot})^4} .$$

P7. How would you answer the question: Why are sunspots dark?

Exercise 14: Parallax Distance Determinations

Learning Objectives

In this experiment, you will make two distance measurements using the method of parallax. In the first measurement, you will use a surveyor's transit to determine the distance to a stake. A second distance determination will give the moon's distance using a pair of photographs of the moon taken at different locations on the earth.

Introduction

The distances to celestial objects is of fundamental importance to astronomy. Is a star in the night sky bright because it is nearby, or is it bright because the star is intrinsically luminous? Knowing the distance to the star answers the question. Photographs of the night sky are only two-dimensional images of space, which has a vast third dimension; distance determinations add that missing third dimension.

Astronomical objects, however, are exceedingly distant, almost to the point of being unimaginable. Who can truly comprehend a light year? Astronomical distances are very large, at least in terms of human scales. So how can such large distances be measured? Traditional devices such as meter sticks and tape measures are obviously inadequate; light-year-long tape measures are not normally available as hardware stores do not stock tapes longer than 100 ft. In astronomy, other less direct methods of obtaining distance must be employed. And because the distances to be measured are large and the methods are not direct, distance determinations are subject to relatively large uncertainties. Consequently, whenever possible more than one method to measure a distance is utilized.

You probably already have some familiarity with the method of parallax that you will use in this lab. Extend your arm about halfway and hold your index finger about a foot or so in front of your eyes. Close your left eye and view your finger from your right eye, noting its position against a more distant background. Close your right eye and now view your finger with your left eye. You will note that its position against the background has shifted. Your finger has not moved, but because you have viewed it from different points (different eyes), it appears to have shifted because of the differing perspectives.

A small amount of experimentation will show that the amount of the shift becomes less the further away you hold your finger and increases the closer you hold your finger to your eyes. Figure 14.1a and b illustrate this idea.

The angle θ represents the shift in the apparent position of the finger. We call $\theta/2$ the **parallax angle**. The distance between the points of observations, i.e., the distance between your eyes, is called the **baseline**.

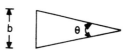

P1. How does the length of the baseline affect the parallax angle?

Figure 14.1 Parallax of a finger when viewed alternately through one eye, then the other.

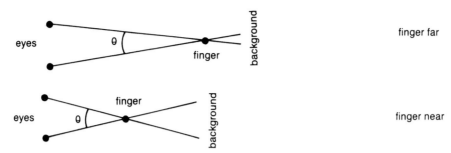

Knowing the length of the baseline and angle θ, the distance to your finger can be calculated (if you have held your finger directly in front of you, and not off to one side). There is a bit of ambiguity as to what we mean by the "distance to your finger." Is it the distance from one eye to the finger, or from the finger to the middle of the baseline? Or suppose you did not hold your finger directly in front of you, but off to one side instead, what is meant by the distance to your finger now?

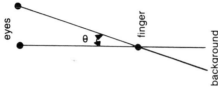

We will see that in astronomical situations, the numerical values of the distances determined in the different ways we mentioned do not differ substantially from each other.

Parallax distance determinations were first employed in astronomy in determining the distance to the moon. Two widely separated observers, say in Denmark and in Italy, would at some particular time note the moon's position relative to the background stars. (No telescope is needed.) The different observers would see the moon projected in different places against the background stars. The size of the moon's parallax can be measured with some ease in this way. (You will make analogous measurements later.)

The *lack* of an observable parallax of a comet proved to be of great importance in the development of modern astronomy. The parallax measurement was attempted by Tycho Brahe (1546–1601), a meticulous astronomer who conducted his observations just prior to the invention of the telescope. With his artificial nose of gold and copper, entourage of dwarfs, legendary drinking parties, and palatial island observatory, he was not the typical astronomer. His competence as a careful observer, though, was known throughout Europe.

The world view prior to the time of Tycho Brahe was that the cosmos was divided into two realms. One was the heavens, which were "free from disturbance, change, and external influence." The other was the earthly realm, which was subject to change and mutability. The boundary was at the position of the moon. This picture of the cosmos was posited by Aristotle (384–322 B.C.). Comets, thought Aristotle, were a kind of atmospheric phenomenon because they would appear and disappear at seemingly random times. Aristotle pictured a comet as an exhalation of hot, dry gas from the earth, perhaps through a fissure, crevice, or volcano. The gas would rise, and when reaching the sky would be heated by the sun and, he thought, burst into flame. Given the limitations of the science of the times, Aristotle's explanation of comets was far from foolish.

In Tycho's time, the unimpeachable authority on comets was still Aristotle. His doctrine that comets were confined to earth's atmosphere because the heavens were fixed and changeless was a cornerstone of the sixteenth-century model of the universe, endorsed by secular and religious authorities alike. This was not a controversial issue. All knowledgeable experts agreed with Aristotle. The first real doubts were raised on a night in 1572 when Tycho looked up at the constellation Cassiopeia and saw "a star brighter than Venus" where no star had been before.

The implications of the appearance of this new star were astounding. How could a "new" star appear in the heavenly realm, which was supposedly changeless and immutable?

Five years later, a great comet blazed across the skies of Europe to overturn decisively Aristotle's now-tottering world view. Because the comet of 1577 was visible for an extended period, Tycho and his colleagues throughout the Continent were able to share information and test each other's hypotheses. The supernova of 1572 prompted Tycho to approach the comet as if it were an astronomical object, rather than an atmospheric disturbance.

Tycho realized that the method of parallax could be applied to the comet, provided it could be observed from at least two widely separated observatories. If the comet is close to the earth, perhaps in the atmosphere, then the perspective will change greatly between the two observatories, and each observer will see the comet in a different position as projected against the background stars. But if the comet is far from the earth, then both observatories will see the comet in the same position relative to the stars. Using parallax, it might even be possible to measure the distance of the comet from the surface of the earth.

Tycho urgently organized and correlated observations of the comet from all over Europe. Tycho himself made many meticulous measurements. If the comet had been within the earth's atmosphere, or even as far away as the moon, a sizable parallax would have been detected. Tycho was able to find no significant parallax. With his precision of measurements, the comet of 1577 had to be much further from the earth than is the moon. The comet must, therefore, be somewhere up there among the planets and stars. Combining international cooperation, elementary mathematics, and simple observations, Tycho found that the conventional wisdom about comets had been dead wrong for two millenia.

The Experiment

Part 1 Distance Determination by Parallax

In the first part of this experiment, you will determine the distance to a stake. The geometry of the arrangement is shown in fig. 14.2 on the next page.

** Show that $\theta = a + b$

Set up your transit at a convenient location (call it point A) so that the stake and some distant reference point are both visible. The reference point should be as distant as practical and should be a *point*, not just some object, but a particular point on some object. Drive a reference pole into the ground at a point (call it B) about 10 m away from your transit and roughly perpendicular to the line of sight between the transit and distance reference point.

Directly under the transit bob, place a marker to indicate the transit's position. Measure the angle between the distant reference point and the stake. This corresponds to angle a, for example, in fig. 14.2. You and your partner should each make several determinations of angle a and average the results. Measure also angle c, the angle between the stake and the reference pole at B. (Note that the sum of angles a and c might not be 90°.)

angle a = _____
angle c = _____

Set up the transit at point B. (You will have to remove the reference pole.) Make several determinations of the angle corresponding to angle b.

With a tape measure, accurately determine the length of the baseline.

angle b = _____
length of baseline = _____

P2. As we have discussed previously, the distance to the stake is a somewhat ambiguous term. Consequently, use trigonometry to obtain the following distances illustrated on fig. 14.2:

1. $\overline{AS}$ = _____ 2. $\overline{BS}$ = _____ 3. $\overline{MS}$ = _____

The following trigonometric relations will probably prove useful:

$$c^2 = a^2 + b^2 - 2ab \cos C$$
$$a^2 = b^2 + c^2 - 2bc \cos A$$
$$b^2 = a^2 + c^2 - 2ac \cos B$$

$$\frac{a}{\sin A} = \frac{b}{\sin B} = \frac{c}{\sin C}$$

Figure 14.2 Geometry for a distance determination to a stake.

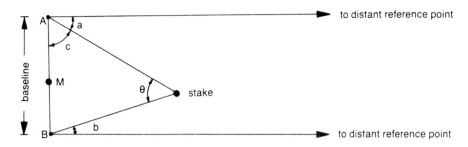

P3. By what percentage do the distances $\overline{AS}$, $\overline{BS}$, and $\overline{MS}$ differ? Which distance would you quote if asked "What is the distance to the stake"?

Part 2 Distance to the Moon

In the second portion of this experiment, you will determine the distance to the moon using the same method of parallax used in the previous part. Figure 14.3a and b are photographs of the moon and Venus taken at the same instant (19 April 1988, 20:15:15 EDT). Two different types of telescopes were used giving different image sizes on the original negatives. The two photographs, however, are reproduced to the same scale, which is indicated in the caption to fig. 14.3a and b. The two stations that took the photographs were 238.8 km apart. At the time the photographs were taken, Venus and the moon were about 45° past the meridian.

To determine the parallactic shift, you will need to superimpose the two photographs. Remove them from the text and separate the two photographs. In the last portion of this experiment, you measured the shift of a nearby object relative to a more distant one. In this part of the experiment, it is easier to measure the shift of the more distant object relative to the nearer one. Thus, you can align the images of the moon and measure the shift in Venus' position. This can be done by superimposing the two photographs over a bright light source or putting pin holes at the cusps of the moon and aligning the pin holes. The shift in the position of Venus can then be measured using the scale of the photographs given in the caption to fig. 14.3.

Determine the distance to the moon.

$$D_{moon} = \underline{}$$

P4. Is the distance to the moon that you determined the earth-moon distance quoted in textbooks? Between what two points is the earth-moon distance measured that is quoted in texts? Between what two points is your distance measured?

Figure 14.3a and b Two photographs of the moon and Venus taken at 20:15:15 EDT on 19 April 1988. The photograph in fig. 14.3a was taken by O. T. Anderson and the author, and the photograph in fig. 14.3b was taken by William Gnirrep. The two stations that took the photographs are separated by 238.8 km. The scale of the photographs is 0.0093 degrees/mm.

Exercise 15
Emission Spectrum of Hydrogen

Learning Objectives
In this experiment, you will calibrate a spectrometer and subsequently use it to measure the wavelengths of the photons emitted in some transitions in hydrogen atoms.

Introduction
Astronomy is a tough business. In many of the other sciences, the objects studied can be examined closely or scrutinized. They can be handled or dissected. Experiments with these objects can be performed over and over again. In large measure, the science of astronomy does not enjoy these luxuries. Apart from the enormously expensive task of launching a spacecraft to retrieve bits of the universe or waiting for some debris to find its way to earth, astronomy primarily advances by people looking at the universe, with the observer and object observed usually separated by distances that are quite large on a human scale.

However, a wealth of information about an astronomical object can be obtained from an analysis of the light, the electromagnetic radiation, that is emitted by the object. Many different physical processes can cause the emission of electromagnetic waves, but the characteristic light emitted (or absorbed) by single atoms is of particular importance in astronomy since the light can be used to identify the atoms that are involved in the emission (or absorption) process.

Atoms can exist only in particular energy states. For example, a fictitious atom A of an element may be found with an energy E_1, E_2, E_3, or E_4 as illustrated below in an energy level diagram. Atom A can have only one of the four possible energies permitted for atoms like itself.

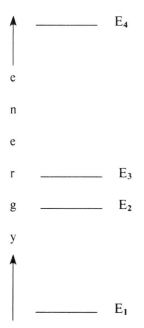

Suppose atom A is in the third energy level, i.e., it has energy E_3. As we have said, in addition to energy E_3, atom A is also permitted to have energies E_1, E_2, or E_4, but for our example it has energy E_3. Now suppose that something makes atom A go from the third energy level to the first energy level. Atom A may make this *transition* spontaneously or it may have interacted with some other atom that induced the transition. Atom A started in level three with energy E_3 and ended in level one with energy E_1. Since E_3 is larger than E_1 and since energy is conserved, the difference in energy must appear somewhere. The energy difference corresponds to the energy of the light that is emitted from Atom A. The bit of light that is emitted in such a process is called a **photon.**

The energy of the light that is emitted (emitted because energy must have left the atom since $E_1 < E_3$) is related to the wavelength λ_{31} of the light by the relation

$$E_3 - E_1 = \frac{hc}{\lambda_{31}}$$

or more generally

$$\Delta E = hc/\lambda$$

where h = Planck's constant = 6.6256×10^{-34} J/s

c = speed of light = 2.998×10^8 m/s

λ_{31} indicates the wavelength of the photon emitted when the atom makes a transition from level three to level one. and ΔE represents the difference in energy between the two levels involved in the transition.

Alternately, if a bit of light of wavelength

$$\lambda = \frac{hc}{E_3 - E_1}$$

were to come by atom A while it was in energy level one, the atom would absorb the light and would make a transition up to energy level three. If the light had a wavelength a bit smaller than this λ or a bit larger than this λ, the transition to level three would not occur.

**Using the energy level diagram for atom A, count the number of different wavelengths of photons that can be emitted from atom A. (Ans: 6) Write down expressions for the wavelengths of the photons emitted in terms of h, c, and the energy of the levels involved in the transitions.

Notice that a large energy difference corresponds to a small wavelength, and, conversely, a small energy difference corresponds to a large wavelength. In the energy level diagram for atom A, the transition 3 → 2 corresponds to a smaller energy difference than 3 → 1 (or 4 → 3 and others). The six different transitions produce six different photons. If we produced a system of many different atoms like A (with the same energy level structure) and somehow randomly excited them so that atoms can be found in all energy levels, then all six transitions would be represented and all six different photons can be observed from the system of A-like atoms. Due to the discrete nature of the energy levels, six and only six different kinds of photons are emitted by A-like atoms. Thus, if you were to analyze the light from the system of A-like atoms you would find only those six wavelengths of photons represented.

A spectroscope can separate light from some source into its component colors or wavelengths in a way similar to the way in which a prism acts on white light. In the example above, the six wavelengths would appear at six different positions. The spectroscope forms a **spectrum** of the light emitted by the A-like atoms or any source of light.

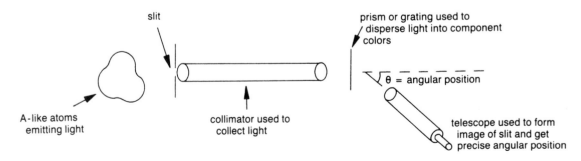

At six different (angular) positions θ, a different wavelength of light would be centered in the telescope. The spectrum of the A-like atoms might look like the following:

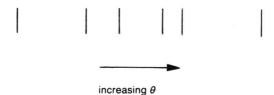

increasing θ

In astronomy, the measurement of the wavelengths of photons emitted when atoms make transitions between energy levels is probably the most important experimental method for learning about the physics of the stars. The diffraction grating spectrometer you will use is an instrument that can give accuracies of up to four significant figures in a wavelength determination. In this experiment, you will use a spectrometer to measure the wavelengths of the photons emitted by excited hydrogen atoms when they make transitions to lower energy levels.

The Experiment

Calibration of the Spectrometer: The optics of the spectrometer should already be adjusted so that images of the slit and reticle crosshairs are both in sharp focus. If the optics have been misaligned, ask the lab instructor to readjust the spectrometer.

The diffraction grating used in the spectrometer is a piece of plastic on which grooves are uniformly spaced a distance d apart. The grating functions like a prism, dispersing light into its component colors.

The angle at which a line of wavelength λ appears is related to the separation d between the grating lines by

$$\lambda = d \sin\theta,$$

provided that the incident light strikes the grating at normal incidence, i.e., perpendicular to the surface of the grating. In order to adjust for normal incidence, rotate the grating stage by very small amounts until a particular spectral line, for example, the green line of mercury, appears at the same angle on either side of the straight-through position.

In order to calibrate the spectrometer, that is, correlate the wavelength of light to the angle at which it appears in the spectrum, someone by some independent means had to measure the wavelength of one spectral color. This has been done for the green line in the mercury spectrum: $\lambda = 546.1 \times 10^{-9}$ m. Measure the diffraction angle θ of the mercury green line, and use its known wavelength of 546.1 nanometers to find the constant d. Make several determinations of d and average the results.

d = _____

The Hydrogen Spectrum: Replace the mercury tube with a hydrogen tube and measure the wavelength of the violet, green, and red lines. For each color, use the average value of the two angles on either side of the straight-through position for the value of θ.

P1. Determine the wavelengths of the three most prominent hydrogen lines.

λ violet = _____
λ green = _____
λ red = _____

The results of a quantum mechanical analysis of the energy levels of a hydrogen atom show that the energy levels occur at values given by

$$E_m = -\frac{k}{m^2}$$

where k is a constant and m = 1, 2, 3, . . . ∞. For the spectral lines that you are observing, the atoms are initially in levels 3, 4, or 5 and all end in level 2. Associate the appropriate pair of numbers (3,2), (4,2), and (5,2) with each wavelength.

P2. Using the relation that the energy of a photon is given by

$$E_{photon} = hc/\lambda$$

calculate a value of the constant k for each transition and average your results.

P3. The energy level of the hydrogen atom is related to the distance from the nucleus where the outer electron is most likely to be found: the higher the energy level of the atom, the further from the nucleus is the electron. What wavelength of light would ionize a hydrogen atom if the hydrogen atom were initially in the first energy level? (In ionizing a hydrogen atom, the outer electron is removed completely from the nucleus.)

Exercise 16
Interactions of Matter and Radiation

Learning Objectives

In this experiment, you will simulate the transport of radiation by the processes of absorption and reemission and see how energy gets transported from the center of a star to its surface. You will also simulate the scattering of radiation by atoms and qualitatively analyze the characteristics of the scattered and transmitted light.

Introduction

At the beginning of the twentieth century, physicists were faced with a dilemma regarding the nature of light. Light seemed to be exhibiting contradictory properties in different kinds of experiments, acting as a wave in some cases and as a collection of tiny particles in others. The basis for the dilemma arose in the latter part of the seventeenth century through the work of two of the most prominent scientists of the time. Christiaan Huygens, who made many pioneering discoveries in optics and the behavior of light, was able to explain much of the behavior of light (as was known at the time) by viewing it as consisting of waves, much like the waves that propagate on a water surface or along a taut string. Isaac Newton, on the other hand, believed that light consisted of a stream of tiny particles, citing the lack of evidence for certain expected wave properties. Newton's idea prevailed at the time, largely due to his greater status as a scientist.

In the very early 1800s, Thomas Young performed a famous experiment in which two beams of light were found to interfere with each other, as though the beams consisted of crests and troughs that would add or subtract when combined, just as the water waves from two pebbles dropped into still water would. Young was even able to measure the wavelengths of light. But by the very late 1800s, it was becoming clear that when light interacted with matter it did so as a particle. In 1905, Albert Einstein showed that many disparate phenomena involving the interaction of matter and radiation could be understood if light carried energy only in discrete amounts, as though light consisted of tiny bullets, each with a specific energy. Soon other experiments seemingly supported the particle nature of light and apparently refuted its wave nature.

Arising from this contradictory set of evidence was the realization that both views were right, and both were wrong; an entirely new picture was needed.

A related digression might be in order: in the original Star Trek series, one episode has particular relevance. In the show "The City on the Edge of Forever," Kirk, Spock, McCoy, Uhura, and a few expendables beam down to the surface of a planet from where Spock has traced the emission of "ripples in time." When they beam down to the planet's surface near the origin of the ripples, they find the decimated ruins of an ancient city (which had obvious classical Greek influences). Near the edge of the ruins lies partly buried an arch-shaped object. Spock tells Kirk that the object is the source of the ripples in time, and with Spock's tricorder merrily whistling in the background, Kirk asks Spock the obvious question: "What is it?"

The object lights up! "A question!," interjects the object in a deep male voice! "A question! I have waited uncounted millennia for a question." Kirk and Spock are surprised that a semi-circular hunk of rock would talk, but they nonetheless talk back. They find out that the object calls itself the "Guardian of Forever." Spock then asks, "Are you being or machine?" The Guardian responds, "I am both and neither." Irritated, Spock replies, "I see no reason to answer in riddles." Whereupon the Guardian says that the answer it gave was the best it could do given the limited intelligence of Spock, irritating the Vulcan further (and Spock was not supposed to have human emotions, ha!).

The point is, humans have mental models of what a being is and what a machine is and the models are mutually exclusive. In the same way, we have models of waves with their energy and location spread out over the entire wavefront and particles with these properties localized. As we examine the universe further we find something that is both wave and particle, and neither—it is the phenomenon of light—and, on Star Trek at least, we find something that is both being and machine, and neither—the Guardian of Forever.

By way of summary we can offer the following general guidelines: When describing how the energy of light propagates, how it goes from here to there, a wave description seems to be appropriate, but when describing how light and other electromagnetic radiation interact with matter, a particle picture is more useful. The physical theory that adequately (and successfully so far) describes this dual nature of light is called **the quantum theory.**

Out of the variety of evidence and experiment came the concept of the **photon**. A photon is thought of as a quantum (particle?) of light that has a characteristic **wavelength** associated with it. (The wavelength is *not* the length of a photon.) The wavelength and the amount of energy associated with the photon are intimately linked; in general terms, the longer the wavelength, the lower the energy. Thus, a photon of red light is less energetic than a blue light photon. Quantitatively, the energy of a photon is given by

$$E = \frac{hc}{\lambda}$$

where h = Planck's constant = 6.6×10^{-34} J·s,
c = speed of light in vacuum = 3.0×10^8 m/s
and λ = the wavelength of the photon.

It is important to understand the fact that a photon carries a precise amount of energy, not some arbitrary or random quantity, and that when a light strikes a surface, this energy arrives in discrete bundles like bullets, rather than in a steady stream. When a photon is absorbed, by an atom say, the energy given up by the photon contributes to the total energy of the atom.

The Experiment

Part 1 Energy Transport

In this part of the experiment, you will simulate the transport of energy from the center of a star to its surface. If you look at energy transport in terms of photons going from here to there, then the flow of energy from the center to the surface is not smooth because each photon of radiation can interact with the atoms and ions it encounters. The energy of a photon is reemitted or redirected by the atom it encounters, usually appearing as a new photon whose direction is randomly oriented. Thus, if you look at successive exchanges of energy by photons, the direction the energy follows becomes a very crooked line, a three-dimensional random walk, as the radiation is repeatedly absorbed, reradiated, and scattered on its way from the center to the surface.

Only four classes of interactions between matter and radiation serve to describe radiative energy transport, but we will neglect details of these processes. For our purposes and though it shall prove to be an oversimplification, the four processes involve a particle (an atom, ion, or electron) absorbing a photon and sooner or later reradiating energy in the form of a *different* photon in a random direction. Thus, the direction of photon-flow is random from interaction to interaction.

Figure 16.1 is a two-dimensional representation of a star where each dot represents an atom, ion, or electron. We will assume that a photon of energy is generated at the center dot and will subsequently follow the energy exchanges between the particles in the star. We will further assume that the energy exchanges occur by means of reradiated photons whose direction is randomly determined. Each time a photon reaches a particle, roll a die to randomly select the direction of reradiation. The number that comes up on the die corresponds to the direction specified on the small diagram alongside the two-dimensional representation of a star in fig. 16.1. For example, if you roll a 4, the reradiated photon goes 1 unit straight down.

As you follow the energy from the center to the surface, mark the path of photons on your diagram and count the number of interactions (rolls) required for energy generated at the center to get to the surface. After the energy reaches the surface, roll the die one more time to allow the last photon to freely propagate. Record the total number of rolls of the die below.

After everyone in your lab section has tracked the energy of one photon from the center to the surface, find the class average of the number of rolls required and record that number below.

Number of rolls required in your case _____
Class average of number of rolls _____

Figure 16.1 Two-dimensional representation of a star with particles in a lattice.

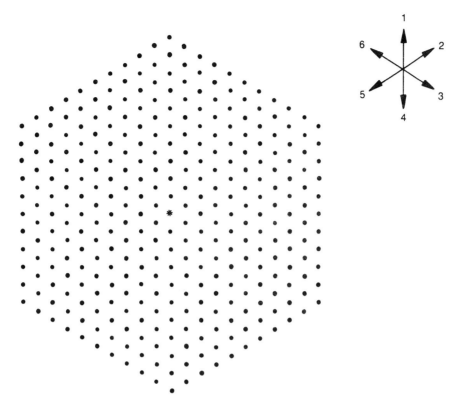

In the optional section at the end of this lab, it is shown that the average number N of rolls required to move through the distance of 10 layers in your diagram is D^2, or 100 rolls.

P1. How does the class average of the number of rolls required compare to the 100 predicted by a random walk analysis?

P2. About 10^{20} interactions occur in getting energy from the sun's center to its surface. If each interaction takes about 10^{-8} sec, how long does it take (in years) for energy to make its way from the core to its surface?

Part 2 Opacity

Atoms exist with only specific discrete energy levels. An atom can jump from one energy level to another rather easily by absorbing or emitting energy corresponding to the difference in energy between the two levels involved in the transition. For example, an atom in allowed energy level E_3 can jump to a higher allowed energy level E_7 by absorbing a photon of energy $E_7 - E_3$. An atom, on the other hand, can make a transition from level E_7 to level E_3 by emitting a photon of energy $E_7 - E_3$. Since the energy of a photon is related to its wavelength, only a specific wavelength will initiate the transition (or result from it). To be specific, a photon of wavelength $\lambda = 656$ nm can be absorbed by a hydrogen atom in its second allowed energy level. That wavelength photon when absorbed will put the hydrogen atom in the third allowed

Figure 16.2 To select randomly the direction of reradiation, roll a die and use the small diagram in the upper right of fig. 16.2. We will assume that any photon leaving the cloud in the '2' direction will reach our telescope on earth; photons leaving the cloud in any other direction do not reach us.

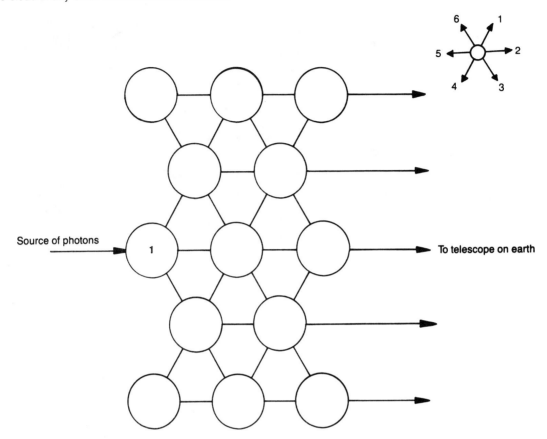

energy level. If the photon had a smaller wavelength or a larger wavelength, the transition would not be initiated and the photon would be much less likely to interact with the hydrogen atom in some way. If a hydrogen atom in the second allowed energy level absorbs a 656 nm photon, the atom is excited to the third energy level, and later (in about 10^{-8} seconds), the hydrogen atom in the third level may spontaneously jump back down to the second level and reemit a 656 nm photon in a completely random direction. (There is also a possibility that the hydrogen atom in the third level drops down to the first allowed level, emitting an ultraviolet photon of wavelength 102 nm, but we neglect this.)

In this part of the exercise, you will simulate the interactions between gas atoms and radiation that can initiate energy level transitions in the gas atoms. We will imagine a source of 656 nm photons and a cloud of hydrogen gas. The cloud could be a nebula near a star or it could be the atmosphere of the star itself. In any case, we wish to examine how the light is affected as it propagates through the gas.

Figure 16.2 is a two-dimensional representation of a cloud of hydrogen gas. Light from the star enters the cloud from the left and strikes the atom numbered 1. We will assume that each photon interacts with every atom it encounters.

****If the incident photons have wavelength 656 nm, which energy level must all the hydrogen atoms be in if each photon interacts with every atom it encounters?

When a photon encounters an atom, it is absorbed and shortly thereafter it is reradiated, but in a random direction. Using a washer as a photon, try to move 10 photons through the cloud.

Number of photons reaching telescope   _____
Number of photons not reaching telescope _____
Percentage reaching telescope   _____
Percentage not reaching telescope   _____

P3. Would you describe the hydrogen cloud as transparent or opaque to radiation of wavelength 656 nm?

Below is a representation of a hypothetical (and strange) star illuminating a cloud of hydrogen gas. The star is strange because it emits equal numbers of photons at wavelengths of 626, 636, 646, 656, 666, and 676 nm. Photons with wavelengths other than 656 do not interact with the gas atoms and continue along their way.

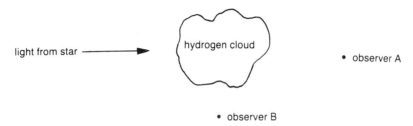

On the graph below, plot the intensity of light as a function of wavelength as seen by observer A. A qualitative sketch is all that is needed, but you can be guided by your results above.

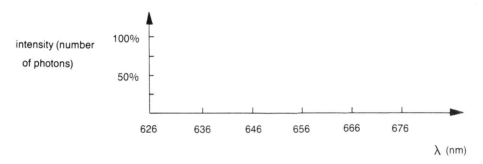

Connect the points by a smooth curve. This represents the *spectrum* of the star.

On the graph below, plot the intensity of light as a function of wavelength as seen by observer B. Again, a qualitative sketch will suffice.

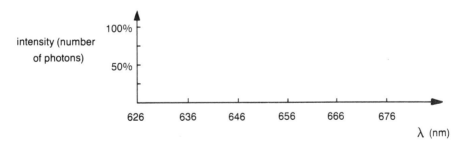

Which observer sees an absorption, or dark line, spectrum? _____
Which observer sees an emission, or bright line, spectrum? _____

Part 3 Scattering Effects

If a cloud of gas has electromagnetic radiation propagating through it, only photons with the particular wavelengths that can initiate transitions in the gas atoms will have a high probability of interacting strongly with the gas atoms, usually by being absorbed. Photons with other wavelengths are much less likely to interact with the gas atoms, but when they do they are usually scattered, that is, their direction of propagation is altered with perhaps a small change in their energies.

Figure 16.3 Illustration of spinner used in scattering analysis. A photon of a particular color (B or R) is scattered only if the spinner points to that color.

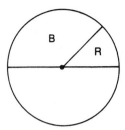

In this part of the experiment, you will investigate qualitatively what differences arise when two different wavelength photons interact with a gas. For most cases, the shorter the wavelength of the photon the greater the probability that it is scattered. For example, a blue light photon is much more likely to be scattered than a red light photon.

You will use the two dimensional representation of a gas depicted in fig. 16.2. This time you will move 10 blue photons (steel washers) and 10 red photons (brass washers) through the gas. You will use a spinner to randomly determine whether a particular photon is scattered.

Half of the area of the spinner is unlabeled. About 1/8 of the total area is labeled R and 3/8 is labeled B. A photon of a particular color is scattered only if the spinner points to the area labeled with that color. If you are moving a red photon and the spinner lands in the unlabeled area or the "B" area, the red photon is not scattered but continues along the direction it had when getting to the circle in which it is located.

As before, start in circle 1. Move 10 red photons through the gas cloud one at a time. If the spinner indicates that the photon is to be scattered, roll the die to determine the direction of the scattered photons. We will assume, as before, that any photon that leaves the cloud is in the '2' direction and will reach the observer; photons leaving the cloud in any other direction do not reach the observer. Keep track of the number of photons that reach the observer, and the number that do not.

After using 10 red photons, use 10 blue photons.

Number of red photons reaching observer   _____
Number of red photons not reaching observer   _____
Percentage of red photons reaching observer   _____
Percentage of red photons not reaching observer   _____

Number of blue photons reaching observer   _____
Number of blue photons not reaching observer   _____
Percentage of blue photons reaching observer   _____
Percentage of blue photons not reaching observer   _____

P4. Would you describe this cloud as more or less transparent to the radiation than in the case in part 2?

P5. Has the light that reaches the observer been "reddened" or "blued" by the scattering processes?

Why is the sky blue? Air molecules preferentially scatter blue light more than red. The atmosphere depletes a beam of light of its shorter (bluer) wavelengths, which are scattered uniformly through the sky. In any direction you look you see the scattered blue light, and so the entire sky is blue (see fig. 16.4). Light of longer (redder) wavelengths reaches you directly along the line of sight.

Figure 16.4 The blue sky. Air molecules let red light pass through relatively unhindered, but blue light is scattered in all directions. Since blue light scatters all around the air, in every direction, you see blue light—and so a blue sky.

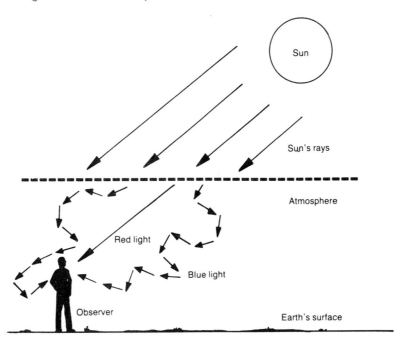

P6. Explain why the setting sun appears red.

P7. Illustrate on the diagram below why the setting sun appears red.

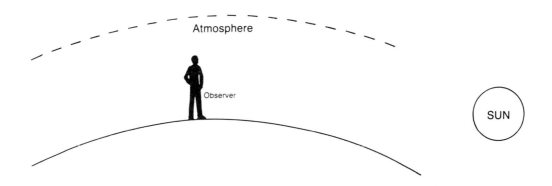

Appendix: The Random Walk (optional)

The transport of radiation as we have modelled it is related to the problem of the "random walk." In its simplest version, we imagine a "game" in which a "player" starts at the point x = 0 and at each "move" is required to take a step *either* forward (toward +x) *or* backward (toward −x). The choice is to be made *randomly,* determined, for example, by the toss of a coin. How shall we describe the resulting motion? In its general form, the problem is related to the motion of atoms (or other particles) in a gas—called **Brownian motion**—and also to the combination of errors in measurements. You will see that the random-walk problem is closely related to energy transport in a star.

Interactions of Matter and Radiation 95

Figure 16.5 The progress made in a random walk. The horizontal coordinate N is the total number of steps taken; the vertical coordinate D(N) is net distance moved from the starting position.

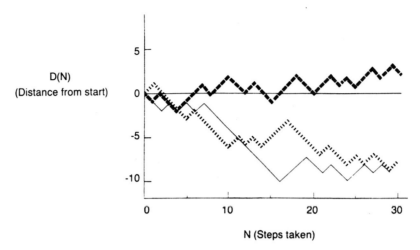

First, let us look at a few examples of a random walk. We may characterize the walker's progress by the net distance D_N traveled in N steps. We show in the graph of fig. 16.5 three examples of the path of a random walker. (We have used for the random sequence of choices the results of the coin tosses.)

What can we say about such a motion? We might first ask: "How far does he get on the average?" Our initial reaction might be that his average progress will be zero, since he is equally likely to go either forward or backward. But we have the feeling that as N increases, he is more likely to have strayed farther from the starting point. We might, therefore, ask what is his average distance traveled in *absolute value,* that is, what is the average of |D|? It is, however, more convenient to deal with another measure of "progress," the square of the distance: D^2 is positive for either positive or negative motion, and is, therefore, a reasonable measure of such random wandering.

We can show that the expected value of D_N^2 is just N, the number of steps taken. By "expected value" we mean the probable value (our best guess), which we can think of as the *expected* average behavior in *many repeated* sequences. We represent such an expected value by $\langle D_N^2 \rangle$, and may refer to it also as the "mean square distance." After one step, D^2 is always $+1$, so we have certainly $\langle D_1^2 \rangle = 1$. (All distances will be measured in terms of a unit of one step. We will not continue to write the units of distance.)

The expected value of D_N^2 for $N > 1$ can be obtained from D_{N-1}. If, after $(N-1)$ steps, we have D_{N-1}, then after N steps we have $D_N = D_{N-1} + 1$ or $D_N = D_{N-1} - 1$. For the squares,

$$D_N^2 = \begin{cases} D_{N-1}^2 + 2D_{N-1} + 1, \\ \quad\quad\text{or} \\ D_{N-1}^2 - 2D_{N-1} + 1. \end{cases}$$

In a number of independent sequences, we expect to obtain each value one-half of the time, so our average expectation is just the average of the two possible values. The expected value of D_N^2 is then $D_{N-1}^2 + 1$. *In general,* we should *expect* for D_{N-1}^2 its "expected value" $\langle D_{N-1}^2 \rangle$ (by definition!). So

$$\langle D_N^2 \rangle = \langle D_{N-1}^2 \rangle + 1.$$

We have already shown that $\langle D_1^2 \rangle = 1$; it follows then that

$$D_N^2 = N,$$

a particularly simple result!

If we wish a number like a distance, rather than a distance squared, to represent the "progress made away from the origin" in a random walk, we can use the "root-mean-square distance," D_{rms}:

$$D_{rms} = \sqrt{\langle D^2 \rangle} = \sqrt{N}.$$

Exercise 17: Observations of δ Cephei

Learning Objectives
In this observing project, you will monitor the brightness variations of δ Cephei to determine its period of variability.

Introduction
For astronomers, determining the distances to the objects they study has always provided a source of major headaches. By the late nineteenth century, distances to a few hundred nearby stars had been painstakingly worked out by the parallax method, which consists of allowing Earth's orbital motion around the sun to alter our perspective, so that relatively nearby stars appear to shift slightly relative to those in the background. But this method worked well only up to a few tens of light years. Beyond that, astronomers knew where stars were in the sky but not where they were in space. Until that third dimension could be supplied, talk of the large scale arrangement of the stars remained only conjecture.

At the end of the eighteenth century, William Herschel, musician and astronomer to George III of England, completed a project to map the heavens with some of the largest telescopes existing at the time. Always ahead of his time, Herschel tried to construct a model of the Milky Way by counting stars in all directions and simply assuming them to be of about the same intrinsic brightness. With this assumption, a faint star would be more distant than a brighter star. Stars, however, come in all sorts of intrinsic brightnesses, and the task defeated him. He ended up with a model of the Milky Way Galaxy resembling a thick botched pancake with the sun at the center. He went on to other matters.

Stuck in two dimensions, astronomers in the closing decades of the nineteenth century shelved most big questions and bent themselves to less inspiring work, compiling star charts and assembling huge catalogues that listed the names, positions, spectra, and colors of stars. Late nineteenth-century university astronomy departments resembled factories of the period, where impervious bosses oversaw rooms full of employees whose repetitious toil threatened their eyesight. But their methods produced results. The columns of dry figures suggested secrets about the stars.

At Harvard University Observatory, the astronomical sweatshops were peopled mostly by women. Edward Charles Pickering hired them, gave them the title "computer," and paid them 25 cents an hour to fill blank catalogue pages with tiny black ink numbers, no mistakes permitted. About 40 women worked as computers for Pickering. Most came and went silently, but a few persisted in trying to make some sense out of the blizzard of data.

One of these "computers," Henrietta Swan Leavitt, gave astronomy its third dimension. Leavitt discovered what is called the **period-luminosity relation** in cepheid variable stars, a way of measuring distances to this particular type of stars. This distance-measuring technique made possible the discovery of the size of the Milky Way system of stars, of the existence of galaxies as vast assemblies of stars external to our own galaxy, and of the expansion of the universe itself.

Leavitt came to Harvard from the Society for the Intercollegiate Instruction of Women (later Radcliffe) in 1895. Pickering stacked photographic plates in front of her and told her to look for variable stars. Leavitt would study two photographs taken of the same area of sky at two different dates and check whether any of the thousands of stars had changed in apparent brightness, betraying itself as a variable.

Variable stars had been observed for over a century and were known to fall into two categories: the **eclipsing binaries** and what might be called **intrinsic variables.** Eclipsing binaries are systems of two stars that happen to be oriented in space so that one star, seen from our viewpoint, periodically passes in front of the other as the two orbit each other. Intrinsic variables primarily are stars that pulsate; their intrinsic brightness varies with time. Some vary irregularly, but others are quite periodic. A few take a year or more per pulse, but others, less leisurely, vary with periods on the order of a month or even a few hours. These more rapid variables (the ones with periods in the range of a few hours to about a month) include the **Cepheids,** so called because the first one identified was δ Cephei, the one you will be studying. The Cepheid variables proved to be the distance indicators astronomers needed.

Henrietta Leavitt examined thousands of Cepheids on the Harvard plates—she discovered 2,400 herself—and as time passed she began to perceive a pattern emerging. The brighter the Cepheid variable star, the longer it took to go through a cycle of brightness variations. She was able to make this important discovery because, as it happened, many of the Cepheids she was assigned to study were in the Large and Small Magellanic Clouds, two satellite galaxies of the Milky Way. Pickering and his staff of computers did not know precisely what the Clouds were, but an important effect operated anyway: All the Cepheids in them were bound up together at roughly the same distance, like fireflies in a bottle far away, so that confusing differences in brightness caused by varying distance were suppressed. Any differences in brightness represented intrinsic differences in luminosity.

The inherent relationship of period to luminosity revealed itself. If the intrinsic brightness, related to the absolute magnitude, of a single Cepheid star could now be determined, distances deep into the cosmos might be measured.

Finding the distance to any single Cepheid proved difficult because not one was near enough to use the method of trigonometric parallax. But several astronomers, by exploiting a statistical method involving the sun's drift among its fellow stars, managed at least to get a fair estimate of the distance to a few short-period Cepheids. Now the distance of any Cepheid variable could be determined from knowing only its average apparent brightness and the period of variability. Astronomy had found its third dimension.

δ Cephei, the star you will be studying, is the prototype of the class of pulsating variables Leavitt investigated and was the first known star of the group. The variability of δ Cephei was discovered in 1784 by the deaf-mute English astronomer John Goodricke, just two years before his death at the age of 21. Goodricke also studied the brightness variations of Algol, and in this case and that of δ Cephei he offered models explaining the brightness variations, which were quite on target.

The Observations

You will be determining the **light curve** (brightness vs. time) of δ Cephei. You begin by locating the constellation Cepheus. Choose a chart from Appendix 2 that gives the relative locations of the constellations for the particular months you will be observing. Finding constellations using a star chart is only initially tricky; once you have found one and see how the size of the constellation in the sky corresponds to the representation on the chart, finding other constellations is much easier. It is a big help to know someone who can point out one or two constellations to get over this initial hurdle. Try working with others in your lab section.

The constellation Cepheus is not prominent, but you can use the nearby constellation Cassiopeia to help point the way. Cassiopeia has a characteristic "W" shape (or "M" shape depending on orientation) and lies along the Milky Way. Immediately to the west are the stars of Cepheus.

Having found Cepheus, use the detailed star chart provided with this lab to find δ Cephei. Note that the ring around the dot representing δ Cephei on the star chart indicates that δ Cephei is a variable star. Next, examine the star chart (and the sky) to find the reference stars listed in table 17.1. Identify as many of them in the sky as you can.

To estimate the brightness of δ Cephei, compare its brightness at any particular time with the brightnesses of the reference stars listed in table 17.1. Each star in table 17.1 has been assigned a letter grade: "A" is the brightest and "I" is the dimmest.

**Why were β Cephei and μ Cephei not used as comparison stars?

Having located δ Cephei and the reference stars, you need to independently observe δ Cephei on as many clear nights as possible. When you have got a good night, try to make observations every one or two hours for as long as is practical. (Your instructor does realize that astronomy is not the only course you are taking.)

For each observation note the time to the nearest minute and assign a letter grade for brightness by comparing δ Cephei with the reference stars.

Tabulate your results on the sheet provided. List the date, time of measurement, the brightness of δ Cephei ("A" to "I"), and sky conditions.

Astronomers measure brightness in magnitudes, an ancient and hardly venerable system. It must be said, however, that the magnitude scale is a logarithmic scale that roughly corresponds to how our eyes respond to light. The apparent magnitudes of the reference stars are provided in table 17.2. Using the data in table 17.2, estimate the maximum and minimum magnitudes of δ Cephei. Remember that brighter objects have numerically lower magnitudes, while fainter objects have numerically higher magnitudes; i.e., a magnitude 1.0 star is brighter than a magnitude 5.0 star.

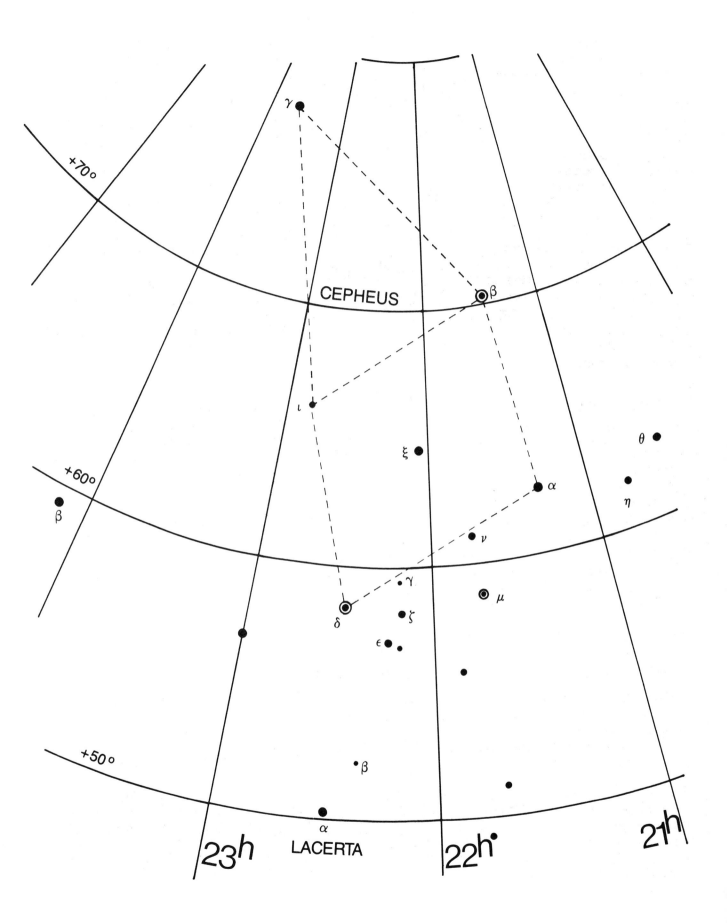

Observations of δ Cephei

Table 17.1 Brightnesses of Reference Stars

Star	Brightness
γ Cep	A
ζ Cep	B
η Cep	B
ι Cep	C
α Lac	D
θ Cep	F
υ Cep	F
β Lac	G
ξ Cep	H
λ Cep	I

Table 17.2 Apparent Magnitudes of Reference Stars

Star	Apparent magnitude
γ Cep	3.2
ζ Cep	3.4
η Cep	3.4
ι Cep	3.6
α Lac	3.9
θ Cep	4.2
υ Cep	4.3
β Lac	4.5
ξ Cep	4.6
λ Cep	5.0

Plot δ Cephei's brightness in magnitudes vs. time to see if you can determine the period P of the pulsations. Your observations are probably very unevenly spaced in time so make sure this unevenness is accurately portrayed in plotting the points for the light curve; that is, *do not* force regularity in the spacing of observations where none exists. Draw a smooth curve through your data. Note that this means that the curve does not have to pass exactly through each data point.

Period of δ Cephei = _____

P1. Below is a modern period-luminosity relation diagram for periodic pulsating variable stars. As you can see, the situation has become a bit more complicated since Leavitt's work. Three different types of periodic pulsating stars are recognized. δ Cephei is a Type I (also called "classical") Cepheid. Type I and Type II Cepheids can be distinguished by their spectra. Using your value of δ Cephei's period and the graph below, determine δ Cephei's average absolute magnitude.

M = _____

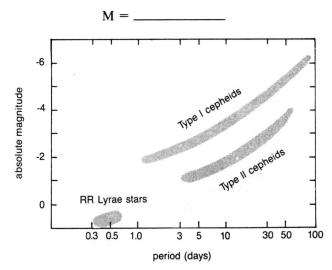

100 Exercise 17

P2. From your observations, deduce δ Cephei's average apparent magnitude, i.e., $(m_{max} + m_{min})/2$.

Average apparent magnitude = m = _____

P3. Using the distance modulus relation

$$m - M = 5 \log_{10} d - 5$$

estimate the distance to δ Cephei. (Note that d in the above equation is in parsecs; 1 pc = 3.26 lightyears.)

d = _____

Observing Log					
No.	Date	Time	Brightness Letter	Magnitude	Sky Conditions
1					
2					
3					
4					
5					
6					
7					
8					
9					
10					
11					
12					
13					
14					
15					
16					
17					
18					
19					
20					
21					
22					
23					
24					
25					
26					
27					
28					
29					
30					

Exercise 18

Algol and Eclipsing Binaries: Observations

Learning Objectives

In this observing project, you will observe how the brightness of Algol changes with time to determine the period of revolution of the components of the Algol system.

Introduction

In our modern world, where electricity has banished darkness from the night so successfully and people live in glittering, illuminated cities and watt-infested homes, we seldom *really* look at the sky on a starry night. But in preindustrial times the sky at night was familiar, if not intimately known. In a passage in *Don Quixote* Sancho Panza tells how a shepherd would know what time it was, and, thus, when dawn was approaching, by observing the position of a particular constellation. If you dip into the works of antiquity all the way back to the dimmest past of any human culture, you will find similar references, a similar appreciation of the skies among people at large. It was not knowledge confined to specialists.

In the constellation of Perseus, one star commanded the attention of the ancient observers. Actually, we really only surmise that the star we call Algol was scrutinized by ancient skywatchers; the inference is based on the names various cultures have given Algol. The name we use is from the Arabic *Al Ra's al Ghul,* "the Demon's Head." It was called *Tseih She,* "the Piled-up Corpses," by the Chinese, and *Rosh ha Satan,* "Satan's head," by the Hebrews. There seems to be a general consensus in its naming.

In the illustration on the next page from a sky Atlas published in 1687, Algol appears as the baleful right eye of the Gorgon Medusa's severed head. Is there any significance to making this identification? Consider this: Algol is unique among all the stars in the sky in that it visibly "blinks" in the course of a single night. This phenomenon was first recorded in the late seventeenth century by Montanari and then by Maraldi. But the most interesting communication about it, reporting a *periodic* pattern of minima and maxima in brightness, was made by the deaf and dumb 18-year-old, John Goodricke. Goodricke reported his observations in a paper to the Royal Society in 1783. Four years later he was admitted to fellowship in the society, and by this time had discovered the variability of two other stars. He had also correctly guessed the source of Algol's variation: Algol is an eclipsing binary.

An eclipsing binary system consists of a bright star and a fainter star in orbit about each other. The plane of this orbit lies along our line of sight from the earth to the system. Consequently, as the two stars orbit each other, one star periodically passes in front of the other. When the fainter star passes in front of and eclipses the brighter star, the total amount of light from the system diminishes. Later in the cycle, the brighter star passes in front of the fainter star and the total light we receive is again diminished, although not as much as when the dimmer, fainter star was blocking the light from the brighter star. By observing the change in brightness of the system during an eclipse cycle, the so-called **light curve,** the orbital period, P, of the binary system can be deduced.

Very precise observations and analysis of the light curve of an eclipsing binary can also yield the size of each star and the inclination of the orbit of the binary system. This information can be combined with spectroscopic data to obtain the mass of each component. For some close binaries, hot spots created when the hotter component shines on the cooler and tidal effects distorting the shapes of the stars from spherical can also be deduced. You can see that eclipsing binaries provide mounds of primary information about stars.

The Observations

All this observing is quite complicated, however. For the present, we will restrict ourselves to determining observationally the light curve (brightness vs. time) for the prototypical eclipsing binary system Algol or β Persei.

You begin by locating the constellation Perseus. Choose one or two charts from Appendix 2 that give the relative locations of the constellations for the particular months you will be observing. Finding constellations using a star chart is only initially tricky; once you have found one and see how the size of the constellation in the sky corresponds to the representation on the chart, finding other constellations is much easier. It is a big help if you know someone who can point out one or two constellations to get over this initial hurdle. Try working with others in your lab section.

Begin by locating the constellation Cassiopeia, a rather prominent grouping of stars. Cassiopeia has a characteristic "W" shape (or "M" depending on orientation) and lies along the Milky Way. Further to the northeast lies the bright star Capella or α Aurigae. Between Cassiopeia and Capella you will find the stars of Perseus.

Having found Perseus, use the detailed star chart provided with this lab in order to find Algol. Note that the ring around the dot representing Algol on the star chart indicates that Algol is a variable star. Next, examine the star chart (and the sky) to find the reference stars listed in table 18.1 (or as many of them as you can).

Table 18.1 Brightness of Reference Stars

Star	Brightness
α Per	A
γ Per	E
δ Per	E
ε Per	D
ζ Per	D
η Per	G
κ Per	G
ο Per	G
α And	B
β Tri	E
α Ari	B

Table 18.2 Apparent Magnitudes of Reference Stars

Star	Apparent magnitude
α Per	1.8
γ Per	3.0
δ Per	3.0
ε Per	2.9
ζ Per	2.8
η Per	3.8
κ Per	3.9
ο Per	3.8
α And	2.3
β Tri	3.0
α Ari	2.0

To estimate the brightness of Algol, compare its brightness at any particular time with the brightnesses of the reference stars listed in table 18.1. Each reference star in table 18.1 has been assigned a letter grade: "A" is the brightest and "G" the dimmest.

**Why was ρ Persei also not used as a reference star?

Having located Algol and the reference stars, you need to *independently* observe Algol on as many clear nights as possible. When you have got a good night, try to make observations every one to two hours for as long as is practical. (Your instructor does realize that astronomy is not the only course you are taking.)

For each observation note the time to the nearest minute and assign Algol a letter grade for brightness by comparing Algol with the reference stars.

If you notice that Algol is dimmer than normal, it is probably in the eclipse phase, and you should make brightness measurements every 20–30 minutes.

Tabulate your results by listing the date, time of measurement, the brightness of Algol ("A" to "G"), and the sky conditions.

Astronomers measure brightness in magnitudes, an ancient and hardly venerable system. It must be said, however, that the magnitude scale is a logarithmic scale that roughly corresponds to how our eyes respond to light. The apparent magnitudes of the reference stars are provided in table 18.2. Using the data in table 18.2, estimate the maximum and minimum magnitudes of Algol. Remember that brighter objects have numerically lower magnitudes, while fainter objects have numerically higher magnitudes; i.e., a magnitude 1.0 star is brighter than a magnitude 5.0 star.

Plot Algol's brightness in magnitudes vs. time to see if you can determine the period, P, of the binary system. Your observations are probably very unevenly spaced in time so make sure this unevenness is accurately portrayed in plotting the points for your light curve; that is, *do not* force regularity in the spacing between observations where none exists. Draw a *smooth* curve through your data. Note that this means that the curve does not have to pass exactly through each data point.

Observing Log

No.	Date	Time	Brightness Letter	Magnitude	Sky Conditions
1					
2					
3					
4					
5					
6					
7					
8					
9					
10					
11					
12					
13					
14					
15					
16					
17					
18					
19					
20					
21					
22					
23					
24					
25					
26					
27					
28					
29					
30					

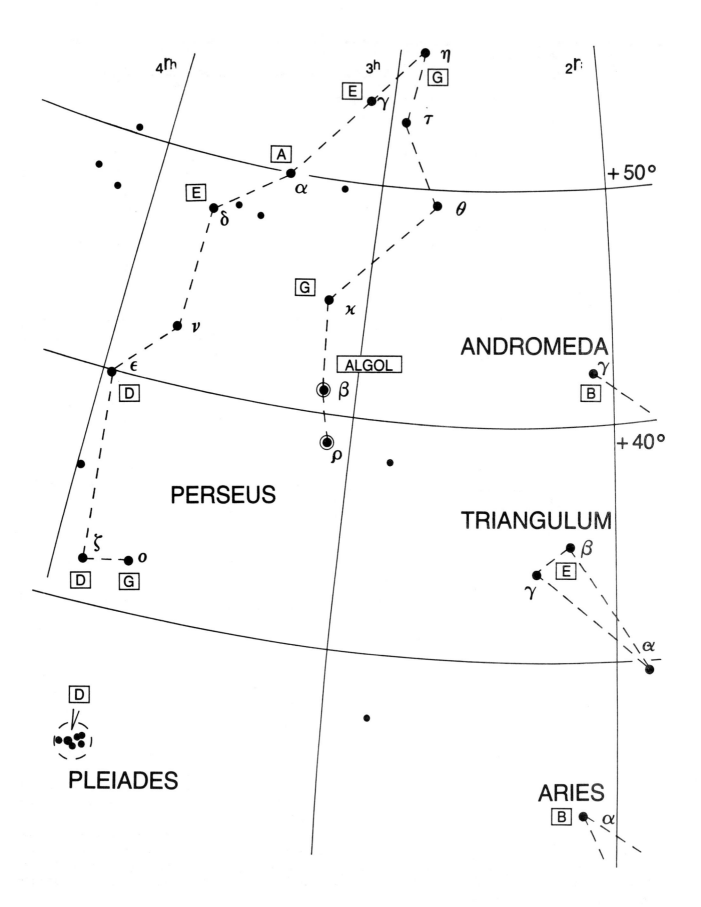

Exercise 19

Algol and Eclipsing Binaries: Simulations

Learning Objectives

In this laboratory exercise, you will graphically construct and subsequently analyze model light curves for eclipsing systems.

Introduction

Stars are born out of immense clouds of interstellar gas and dust. While all the details of the processes involved in the collapse of the clouds to form individual stars are not known, it is clear that a single large cloud gives rise to a cluster of stars, stars gravitationally bound as a group. Tidal forces from the Galaxy and passage near other stars and massive clouds gradually disperse the cluster, but stars that were formed very close together and are strongly bound can remain bound even after the rest of the cluster has lost its identity.

How common are multiple star systems? One measure might be obtained by a survey of the nearby stars. In this way, we can be sure of a fairly complete sample; fainter components of a multiple star system would be less likely to escape detection. Of the 33 nearest stars (including the sun), only 16 are single stars; the rest are in multiple star systems.

Multiple star systems, especially binaries, provide the only direct measure of one of the most important astrophysical properties of a star—its mass. To more than just a first approximation, stars have the same composition but they differ mostly in mass. A star's mass is the principal parameter that determines its energy output, and consequently, its temperature and spectrum, and its evolution and final stable endpoint. Without overstating the point at all, mass is the single most important parameter that governs the life-course of a star.

Mass is involved in gravitational interactions and so to determine a star's mass we need to examine the gravitational effects of one star on another. Here is one instance where binary star systems provide critical astrophysical information.

We will also see that in a certain class of binary stars, eclipsing systems, we can also deduce the *sizes* of stars. With more than 4,000 eclipsing binaries catalogued, they represent an abundant source of astrophysical information.

Discussion

To simplify our discussion of eclipsing binaries, we will assume that the two components of the system revolve around each other in circular orbits. Figure 19.1 illustrates the details.

The two components of the binary system revolve around their common center of mass, marked by an X in the figure. The two stars are always on opposite sides of the center of mass and revolve around it with the same period.

****Which star in the above figure has the larger mass, the one with the smaller orbit or the one with the larger orbit? Explain.

Also note that the separation of the stars remains the same as the stars orbit, so when viewed from one of the components, say the more massive one, the other component orbits it in a circular path. Frequently, in the following discussion we will be referring to this relative orbit.

Actually the assumption of circular orbits is not so bad. Most stars in binary systems that are seen to eclipse each other are relatively close to each other. Their mutual proximity enhances the effects of tidal interactions and other perturbations, and these effects tend to produce circular or nearly circular orbits.

In an eclipsing binary system, the orbital plane of the system lies directly along our line of sight, or nearly so. During the period of revolution of the eclipsing binary, there are two times where the light from the system diminishes: once when the smaller star passes behind the larger one and is eclipsed, and once when the smaller star passes in front of the larger one and eclipses part of it. If the smaller star goes completely behind the larger one, that eclipse is total and the other eclipse half a period later is annular.

Figure 19.1 Top view of eclipsing system with circular orbits.

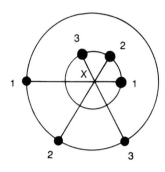

Figure 19.2 Eclipsing binary that produces total eclipses.

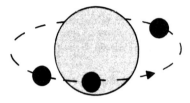

Figure 19.3 Eclipsing binary that produces partial eclipses.

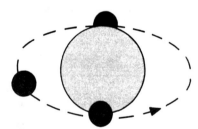

If the smaller star is never completely hidden behind the larger stars, both eclipses are partial.

Each interval during an eclipse when the light from the system is diminished below normal is called a **minimum**. Both minima are not, in general, equally low in intensity. The same area of each star is covered during an eclipse (e.g., if the eclipses are total or annular, an area equal to the cross-sectional area of the smaller star is eclipsed in each minimum), but the relative amount of light intensity drop at each minimum depends on the relative surface brightnesses of the two stars, and hence on their temperatures. **Primary minimum** occurs when the hotter star is eclipsed (whether it is a total, annular, or partial eclipse), and **secondary minimum** occurs when the cooler star is eclipsed. A graph of the light intensity from an eclipsing system as a function of time through a complete period is called the system's **light curve**.

The Experiment

In order to see how astrophysical parameters of a binary system can be deduced, we will simulate the orbital motion of a pair of stars and construct the light curve that results from the motions of the stars. In all of the following, we will work with relative orbits, that is, we will consider the primary star to be central and stationary while the secondary star describes a circular orbit around the primary.

Make a xerox of the three model stars on the following page. Each model star has its surface quadrile ruled with each unit area having a brightness number inscribed. The largest star has 5 units of brightness per unit area, the intermediate size star has 10, and the smallest about 2. Parts of a unit area near the edges of a star have a proportionally

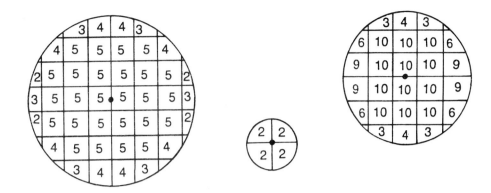

smaller number inscribed. The total brightness of the largest diameter star is 192 units, 200 units for the intermediate size star, and 8 for the smallest.

Place the model star with the largest diameter (5 units of brightness per unit area) at the center of the ellipse in the diagram. The ellipse represents a circular orbit tilted so that it is nearly along our line of sight. Along the periphery of the ellipse are numbered points corresponding to differing *times* during a complete period of revolution. The points are uniformly spaced in time except for the near and far portion of the ellipse. That is, the star takes the same amount of time to go from point 1 to point 2 as to go from 2 to 3, and from 3 to 4, 4 to 5, and 5 to 6. Points are bunched at the near and far portions of the orbit; the time to go from 6 to 7 is one-half of the time to go from 5 to 6. The time interval to go from 7 to 8 is the same as that from 5 to 6. Going from 8 to 9 resumes the standard time interval. The axis on the graph you will be using will automatically keep track of any changes in the time steps.

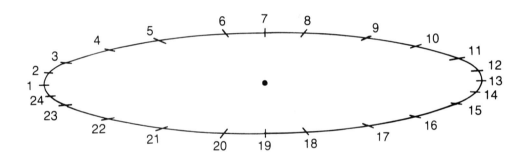

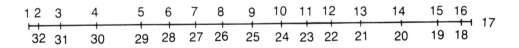

Algol and Eclipsing Binaries: Simulations

Now place the intermediate diameter star with its center on point 1 at the ellipse. At the time corresponding to the star at point 1, determine the brightness of the total system in units of brightness. Record the brightness in table 19.1. Move the star in the succession of points 2, 3, 4, etc., and let us agree that at point 5 the intermediate diameter star begins moving *behind* the large diameter star. At each point, record the total brightness of the system in table 19.1. You will have to compensate for parts of areas obscured, for example, if one-third of 9 unit brightness patch is obscured by a star, then only 6 units will contribute to the total brightness of the system.

Plot your data from table 19.1 on the first graph. Draw a *smooth* curve through the points on the graph.

Table 19.1 Brightness vs. Time for Eclipsing System

Star at Point . . .	Brightness	Star at Point . . .	Brightness
1		13	
2		14	
3		15	
4		16	
5		17	
6		18	
7		19	
8		20	
9		21	
10		22	
11		23	
12		24	

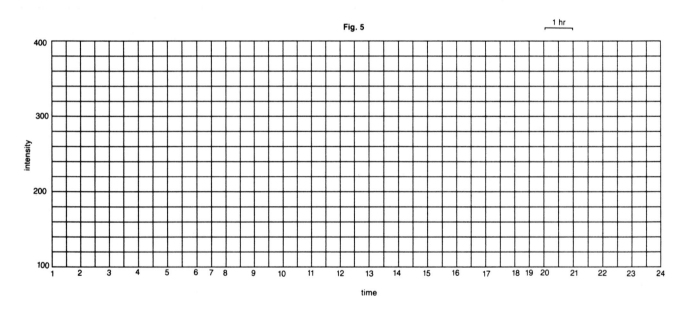

Fig. 5

110   Exercise 19

We will model an eclipsing binary system whose orbital plane is exactly in our line of sight using the numbered line. Place the intermediate diameter star at the center of the line, that is, the center of the intermediate diameter star on point 9 (which is coincident with point 25). Start the smallest star at point 1 and proceed as before, letting the small star pass behind the primary as it goes from points 6 to 10. Record your data for this system in table 19.2, and plot those data on the second graph.

**On the two graphs, which light curve represents an eclipsing system with total eclipses and which with partial eclipses?

Table 19.2 Brightness vs. Time for an Eclipsing System

Star at Point . . .	Brightness	Star at Point . . .	Brightness
1		15	
2		16	
3		17	
4		18	
5		19	
6		20	
7		21	
8		22	
9		23	
10		24	
11		25	
12		26	
13		27	
14		28	

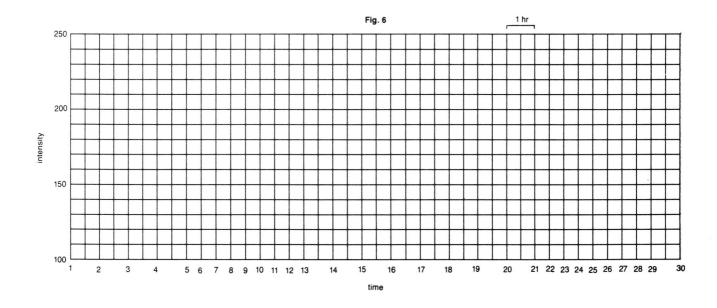

Fig. 6

Algol and Eclipsing Binaries: Simulations

Analysis

We will work primarily with the light curve in the second graph in analyzing eclipsing systems.

Suppose that from a spectroscopic study of our model eclipsing binary, it is found that the relative velocity of the secondary is $v = 330$ km/sec $= 3.3 \times 10^5$ m/s.

Determine the period of the model eclipsing system whose light curve appears on the second graph. Express your determination in both hours and seconds. Calculate the radius of the secondary's circular orbit in meters.

period = _____ hr = _____ sec.
radius of orbit = _____

Analysis of eclipsing binary light curves directly gives one other property of the stars: their diameters. Let us see how.

For simplicity, consider a pair of stars whose orbital plane lies exactly in the line of sight. In this situation as the smaller star orbits the larger, the smaller star moves exactly across a diameter of the larger star. The light intensity of the system begins to diminish at a, when the smaller star starts passing in front of the larger. When it is completely in front of the star at point b the light remains constant, for no more area is blocked off as it moves across. At c, the small star starts to move off from in front of the large star and the light intensity begins to rise, reaching the original level at point d when the larger star is completely uncovered.

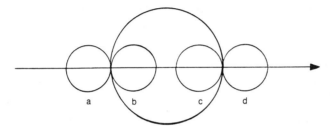

From the diagram, you can see that the smaller star moves its own diameter from point a to point b. It moves the diameter of the larger star in going from point a to point c.

Pl. Using the relative velocity of the small star and your light curve on the second graph, calculate the diameters of both stars.

In 1619, after about 15 years of empirical numerological shennanigans, Johannes Kepler published what was eventually called his Third Law of Planetary Motion, namely:

$$P^2 = kr^3;$$

that is, the square of the orbital period was proportional to the cube of the average distance from the orbit center.

Isaac Newton, after deducing the form of the law of gravitation, was able to generalize Kepler's result and showed that the proportionality constant k had structure and, in fact, depended on the total mass of the system. Specifically,

$$P^2 = \frac{4\pi^2}{G\, M_{total}} r^3$$

or

$$M_{total} = \frac{4\pi^2}{G} \frac{r^3}{P^2},$$

where G is the constant of universal gravitation and $G = 6.67 \times 10^{-11}$ mks units.

P2. Using Newton's generalization of Kepler's Third Law, calculate the total mass of the eclipsing system.

The amount of energy emitted each second for every square meter of a blackbody at temperature T (in kelvins) is

$$E = 5.67 \times 10^{-8} \, T^4 \, W/m^2.$$

Stars radiate energy very much like blackbodies.

From the light curve on the second graph, we can find the ratio of the temperatures of the stars. During both primary and secondary eclipses, the same amount of area is hidden from our view. When the smaller star is in front of the larger, the small star hides an area of the larger star equal to its own cross-section. Then when the small star is eclipsed, the same amount of area is hidden. Any difference in the amount of light lost during the two eclipses must arise from the differences in the temperatures of the two stars, not from the differences in the area hidden. Thus, the relative depths of the two eclipses are related to the ratio of the stars' temperatures.

P3. Calculate the ratio of the temperatures of the two stars whose light curve is on the second graph.

** Will this analysis work in the case of partial eclipses, as on the first graph?

Complications

If the orbits of the stars are not exactly in the line of sight nor perfectly circular, the analysis we have utilized is geometrically more complicated, but the principle is the same. Other complications that can arise are manifested in the light curve of the system. Let us examine some of the complications that can arise and see if we can deduce the effects on the light curves.

1. Close binary. Frequently, the two stars of an eclipsing binary are so close together that they suffer severe tidal distortion and have shapes more like footballs than like spheres.

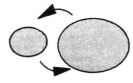

2. Radiative heating. Still another complication arises when the two stars are relatively close together and are very different in temperature. The hotter star can heat up the portion of the cooler star that is nearest it, which causes that portion of the cool star to radiate more intensely than the rest of its surface.

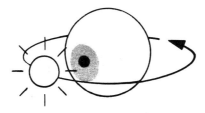

3. Elliptical orbits. Even though orbits of most eclipsing binaries are rather circular, many still have noticeably elliptical orbits.

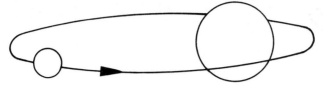

P4. In the three light curves that follow, identify which effect from above, 1, 2, or 3, gave rise to the light curve.

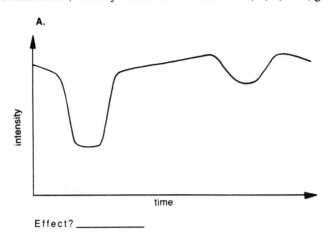

Effect? _____

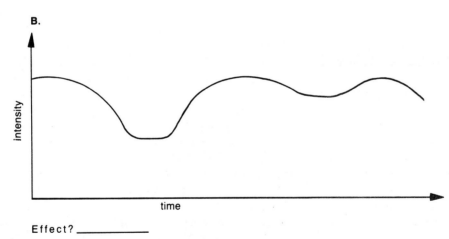

Effect? _____

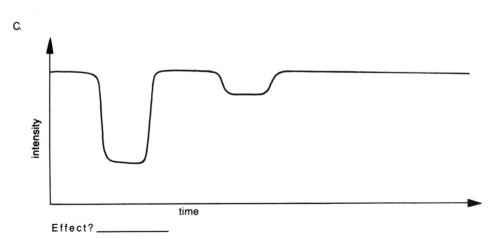

Effect? _____

Exercise 20

Stellar Spectra and the Hertzsprung-Russell Diagram

Learning Objectives

In this exercise, you will obtain some physical properties of stars by examining their spectra. You will also examine the range in fundamental stellar characteristics by analyzing a Hertzsprung-Russell diagram.

Introduction

In 1835, the materialist philosopher Auguste Comte (1798–1857) was searching for an example of a sort of knowledge that would always be hidden. The example for the epitome of the unknowable that he chose was the composition of the distant stars. We could never physically visit them, he thought, and with no sample in hand it seemed we would forever be denied knowledge of their composition.

Only two years after the death of Comte, Gustav Kirchhoff discovered how to divine the composition of the stars. Kirchhoff, a professor of physics at Heidelberg University and whose lectures were noted for their dullness, was interested in the spectra of light that could be obtained by heating various substances and passing the emitted light through a prism. Examples of spectra are illustrated in fig. 20.1. The lines are really images of a single thin slit used to produce the spectra.

Kirchhoff made the discovery that each element produced its own unique pattern of spectral lines, its own fingerprint, so to speak. The spectra of the heated elements consisted of very bright lines. Sodium, for example, had two very bright orange-yellow lines in its spectrum (plus other fainter ones). The sun, however, had a spectrum consisting of a bright background continuum crossed by *dark* lines.

How did the dark lines originate? The bright background continuum is produced by relatively deep layers in the sun. The intensity of any particular wavelength is dependent on the temperature of the material: the hotter the material, the brighter it is at every wavelength. On its way to an observer on earth, the light from the hot, deep layers passes through the relatively cooler upper layers. In these upper layers, particular kinds of atoms, sodium for instance, absorb wavelengths characteristic of the atoms. Soon after absorbing light of particular wavelengths, the atoms will reradiate the light *but in random directions*. The light that was on its way to an observer on earth gets mostly redirected, and consequently particular wavelengths of light are largely absent, hence the appearance of a continuous spectrum crossed by dark lines. (Since the original path to an observer on earth is one of the possible directions for the reradiated light, the dark lines in the sun's spectrum are not completely black.)

The appearance of the dark lines in the solar spectrum associated with the bright lines of sodium gas visible in the laboratory demonstrated that sodium was present in the sun. All Kirchhoff had to do now was to identify the other dark lines in the sun's spectrum—find out which elements had bright lines in that particular place—and he could state with confidence that that element existed in the sun's atmosphere. Eventually, he was able to identify hydrogen, iron, magnesium, calcium, and several other known elements, including gold. Kirchhoff's bank manager asked him one time, "What use is gold in the sun if it cannot be brought down to earth?" Just prior to this Kirchhoff was awarded a prize in gold sovereigns for his scientific work and held up to his banker a portion of the prize, commenting, "Here is some gold from the sun!"

During the solar eclipse of 1868, the astronomer Pierre Jules Cesar Janssen went to India taking a spectroscope with him, and observed the lines of an unidentified element in the solar spectrum. He forwarded his observations to Norman

Figure 20.1 Major types of spectra. This figure has short wavelengths (blue) to the left with wavelength increasing to the right (toward the red). At the top is a continuous spectrum such as you would observe from a glowing solid. Below it is the absorption line spectrum of the sun. Here only the most prominent dark lines are indicated by the element that produces them. Below the sun's spectrum are the bright line spectra of selected elements. Note how the bright lines of sodium and hydrogen line up with the dark lines of the same elements in the sun's spectrum. (Courtesy Eastman Kodak Company.)

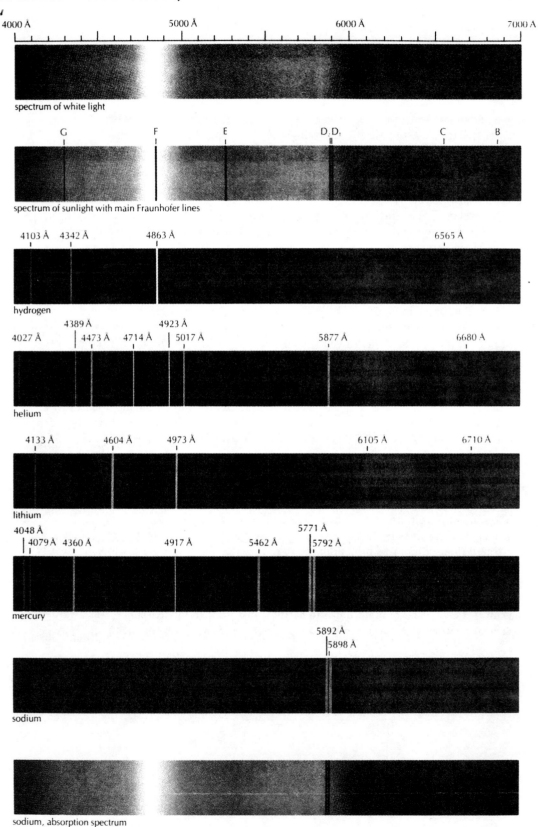

(All wavelengths are measured in vacuum.)

Courtesy Eastman Kodak Company

Figure 20.2 Relative absorption of dark lines of various elements as a function of temperature.

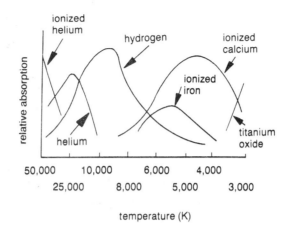

Lockyer, the astronomer who would write about Stonehenge and Egyptian temples, who happened to be a prominent authority on solar phenomena. Lockyer confirmed that these observations represented an unknown element. It was labeled helium, named after the Greek sun god Helios. In 1895, 27 years after having been discovered in the sun, helium was discovered on earth by William Ramsey.

In the early years of the twentieth century when the atomic and molecular processes that produce spectra came to be understood, additional physical properties of the stars could be deduced. Besides composition, physical quantities such as temperature, relative velocity, pressure, and magnetic field intensity could be deduced from a careful analysis of stellar spectra. It seems that astronomical spectroscopy is an almost magical technique. Auguste Comte picked a particularly unfortunate example.

A bewildering array of different spectra presented itself when the spectra of the stars were examined at the turn of the century. In some, hydrogen lines were very prominent. In others, calcium was most noticeable. Did this mean hydrogen was more abundant in one star and calcium in another?

Annie Jump Cannon examined and classified an enormous number of stellar spectra (over 250,000!) in the hopes of discerning some pattern in the spectra. She developed some broad categories and labeled them class A, B, C, and so on. But it was Cecelia Payne, however, who deduced in the late 1920s that the differing spectra of the stars were due principally to the differing *temperatures* of the stars.

Payne's analysis was based on the interaction of photons of electromagnetic radiation with atoms in a distribution of energy states. In brief, a particular kind of atom, hydrogen for instance, is more or less effective at absorbing radiation depending on the temperature of atoms that are doing the absorbing. The effectiveness in absorbing the appropriate wavelengths of radiation for various elements is plotted in fig. 20.2 as a function of temperature.

Let us look at hydrogen in particular since the absorption lines of hydrogen are discernible in the vast majority of stellar spectra. The hydrogen absorption lines that appear in the visible part of a star's spectrum (as contrasted with a star's ultraviolet or radio spectrum, for example) are due to hydrogen atoms initially in the second energy level. Hydrogen has one lower energy level and infinitely many at higher levels. If a star is very cool, 4,000 K for example, there are few collisions energetic or violent enough to excite hydrogen atoms, and most atoms are in their ground state, the lowest energy level. If most atoms are in the ground state, there are few atoms that can absorb photons to produce the absorption lines of hydrogen in the visible part of the spectrum. (These absorption lines are also called **Balmer lines.**) As a result, we expect to find weak Balmer lines in the spectra of cool stars (see fig. 20.3).

In very hot stars, about 40,000 K, the collisions between gas atoms are so energetic that the hydrogen atoms are in energy states *higher* than the second energy level or even ionized. Again, because few hydrogen atoms are in the second level, the Balmer lines are weak.

But at some intermediate temperature, around 10,000 K from figs. 20.2 and 20.3, many hydrogen atoms are in the second level and the Balmer lines are strong.

Figure 20.3 The spectral classification scheme from O (hottest) to M (coolest). The temperature sequence is based on the strengths of the hydrogen Balmer lines and those of other elements such as calcium and iron. (Kitt Peak National Observatory)

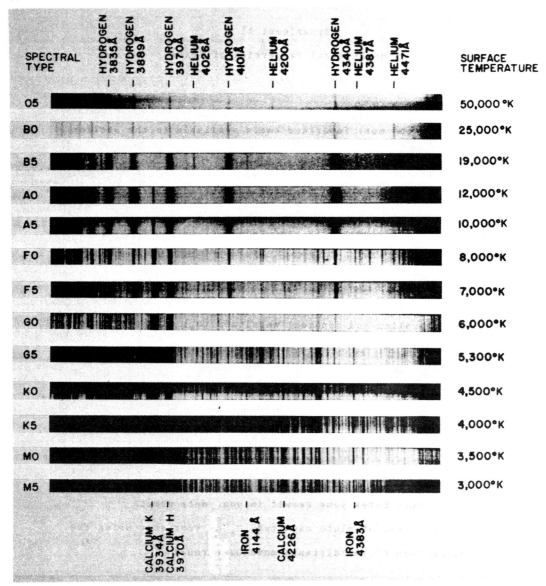

The same process affects the spectral lines of other elements, but the temperature at which they reach maximum strength differs from each element. Together the strengths of the various absorption lines can pinpoint a star's temperature. For example, if we photographed the spectrum of a star and found medium-strength Balmer lines and strong helium lines, we could conclude it had a temperature of about 20,000 K. But if a star had relatively weak Balmer lines and strong lines of ionized iron, we would assign it a temperature of about 5,800 K, similar to the sun.

Eventually, Payne was able to rearrange Cannon's classes into a decreasing temperature sequence that ran O, B, A, F, G, K, M. Figure 20.3 illustrates the results.

Figure 20.4 Examples of stellar spectra from various spectral classes. (Kitt Peak National Observatory)

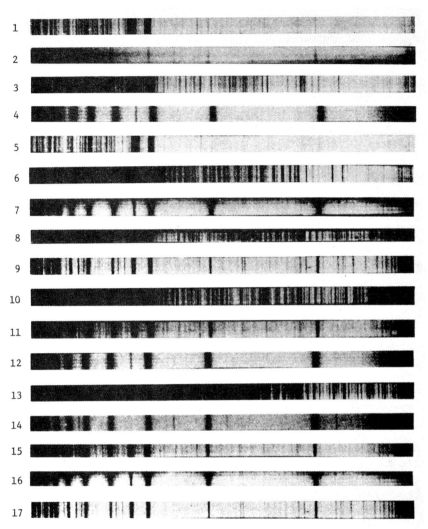

The Experiment

Part 1 Classification of Spectra

Figure 20.4 shows the spectra of 17 different stars. Using the standard spectra in fig. 20.3, classify each of the 17 spectra in fig. 20.4 as to type. In addition, determine the temperature of each of the 17 stars whose spectra appear in fig. 20.4.

P1. The spectra of three stars are illustrated below. Lines *other than* the Balmer lines of hydrogen are identified.

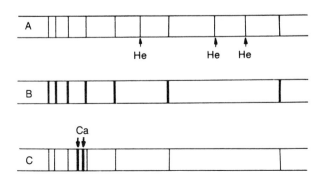

Stellar Spectra and the Hertzsprung-Russell Diagram   119

Use fig. 20.2 to place the stars whose spectra are A, B, and C in order of increasing temperature, i.e., from coolest to hottest. (Assume that the chemical compositions of the stars are identical.)

      Coolest _____
Intermediate _____
      Hottest _____

Let us pick a particular star in fig. 20.4, for example, #12. From a photograph of the star field around the star whose spectrum is #12 in fig. 20.4, we could measure the star's apparent magnitude, 2.4 for instance. Suppose this star is near enough to have its distance determined. The distance to star 12 is 50 parsecs. Knowing the star's apparent magnitude and distance, its *absolute* magnitude can be determined from

$$M = m + 5 - 5 \log_{10} d$$

where d is in parsecs.

**For star #12, verify that $M = -1.1$.

The absolute magnitude of a star is related to its intrinsic luminosity (its power output) by the relation

$$\frac{L_{star}}{L_{sun}} = 10^{-0.4(M_{star} - M_{sun})}$$

where $M_{sun} = +4.84$ and $L_{sun} = 4 \times 10^{26}$ W.

For star 12, $L_{star\ 12}/L_{sun} \approx 240$,
so star 12 is about 240 times more luminous than the sun.

P2. Star #1 has an apparent magnitude of 3.5 and a distance of 6.5 parsecs. Determine its absolute magnitude and its luminosity.

Thus, a star's luminosity can be determined. How do you think a star's luminosity depends on its temperature? We will have to do some checking with real stars. Your guess is probably okay, but a few oddball stars will not behave the way you think.

We will eventually make a plot of stellar luminosity *versus* temperature, but as a prelude let us make a similar plot with human beings. The graph that appears in fig. 20.5 is a plot of the height *versus* weight for a sample of people. Notice the weight axis increases to the *left*.

Figure 20.5 Height vs. weight for a sample of people.

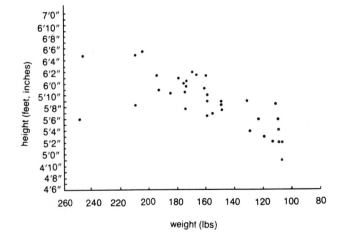

P3. What general statement can you make about the apparent height-weight relationship among human beings?

Part 2 H-R Diagrams

The diagram of height vs. weight for human beings visually sorts people. Other people parameters are hidden in the diagram, parameters related to height and weight. For example, as you probably could have guessed, tall people are generally heavier than shorter people, and they probably have larger shoe sizes, too. Even though shoe size does not appear in the plot, this parameter is correlated to the parameters that are plotted.

In the next part of the experiment, we will make various plots of absolute magnitude vs. spectral class for various samples of stars. (Equivalent to this plot would be one of luminosity vs. temperature or color index.) Even though only absolute magnitude and spectral class appear explicitly as parameters, there are some closely related parameters that do not appear, e.g., mass and radius. In this experiment we will concentrate on plots of absolute magnitude vs. spectral class. Such a graphical sorting of stars is referred to as an H-R diagram after Ejnar Hertzsprung and Henry Norris Russell who did the pioneering work near the turn of the century.

Table 20.1 is a list of the brightest stars visible in the sky. Plot an H-R diagram for those stars on the grid provided.

Table 20.1 The Brightest Stars

Star	m_v	M_v	Type
Sirius	−1.4	1.5	A1
Canopus	−0.7	−4.0	F0
Rigil Kentaurus	−0.3	4.4	G2
Arcturus	−0.1	−0.3	K2
Vega	0.0	0.5	A0
Capella	0.1	0.0	G2
Rigel	0.1	−7.1	B8
Procyon	0.4	2.7	F5
Betelgeuse	0.4	−5.6	M2
Achernar	0.5	−3.0	B5
Hadar	0.6	−3.0	B1
Altair	0.8	2.3	A7
Acrux	0.8	3.9	B1
Aldebaran	0.9	−0.7	K5
Antares	0.9	−3.0	M1
Spica	0.9	−2.0	B1
Pollux	1.2	1.0	K0
Fomalhaut	1.2	2.0	A3
Deneb	1.3	−7.1	A2
Beta Crucis	1.3	−4.6	B0
Regulus	1.4	−0.6	B7
Adhara	1.5	−5.1	B2
Castor	1.6	0.9	A1
Shaula	1.6	−3.3	B1
Bellatrix	1.6	−2.0	B2
Elnath	1.7	−3.2	B7
Miaplacidus	1.7	−0.4	A0
Alnilam	1.7	−6.8	B0

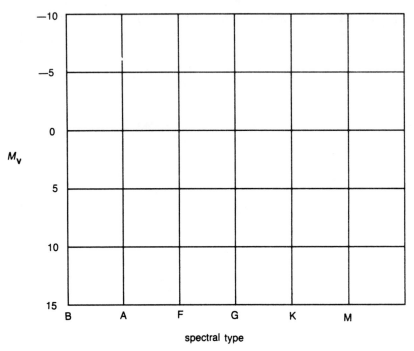

H-R diagram for the brightest stars

Plot the sun on the H-R diagram (M_v = 4.84, Type G2)

P4. How do the brightest stars in the sky compare to the sun in intrinsic brightness or luminosity? Are the bright stars bright because they are near or because they are intrinsically bright?

P5. Capella is the same spectral type as the sun, and consequently has the same surface temperature, but Capella is much more luminous than the sun. How can that be?

Table 20.2 is (what we think) a complete sample of the stars within 13 light years of the sun. Plot another H-R diagram for these stars on the grid provided. Include the sun's position. Two stars happen to be white dwarfs, Sirius A and Procyon B. The spectra of white dwarfs are often difficult to interpret and can only be broadly classed. Sirius A and Procyon B are of spectral type A.

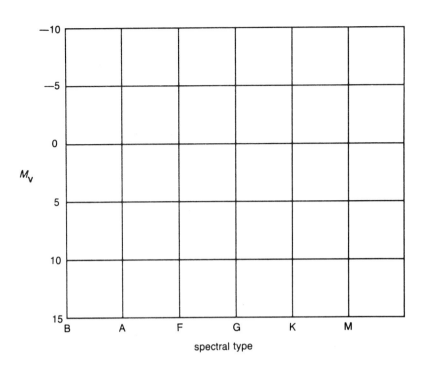

H-R diagram for the nearest stars

Table 20.2

Star	m_v	M_v	Spectral Type
Sun	−26.8	4.8	G2
α Centauri C	0.1	15.4	M5
α Centauri A	1.5	4.4	G2
α Centauri B	11.0	5.8	K5
Barnard's Star	9.5	13.2	M5
Wolf 359	13.5	16.7	M6
Lalande 21185	7.5	10.5	M2
Luyten 726-8A	12.5	15.3	M6
Luyten 726-8B	13.0	15.8	M6
Sirius A	−1.4	1.5	A1
Sirius B	7.2	10.1	wd A
Ross 154	10.6	13.3	M5
ε Eridani	3.7	6.1	K2
Luyten 789-6	12.2	14.6	M6
Ross 128	11.1	13.5	M5
61 Cygni A	5.2	7.5	K5
61 Cygni B	6.0	8.3	K7
ε Indi	4.7	7.0	K5
Procyon A	0.3	2.7	F5
Procyon B	10.8	13.1	wd A
Σ 2398A	8.9	11.2	M4
Σ 2398B	9.7	12.0	M4
Groombridge 34A	8.1	10.4	M1
Groombridge 34B	11.0	13.3	M6
Lacaille 9352	7.4	9.6	M2
τ Ceti	3.5	5.7	G8
Luyten's Star	9.8	11.9	M4
Lacaille 8760	6.7	8.8	M1
Kapteyn's Star	8.8	10.8	M0
Kruger 60A	9.7	11.7	M4
Kruger 60B	11.2	13.2	M6

P6. How does the sun compare in brightness to stars in the solar neighborhood?

P7. Many astronomy texts refer to the sun as an "average" star. In what sense is the sun average? In what sense is it not?

P8. (*Optional*) Suppose you were provided a list of about 30 or so stars in some distant part of our Galaxy. The list included the absolute magnitude and spectral class of the stars on the list. How might the H-R diagram for these stars differ from the one generated from table 2? Be specific as to what part of the Galaxy you are considering.

Exercise 21

Distances to Cepheid Variable Stars

Learning Objectives

In order to utilize Cepheid variable stars as distance indicators, you will establish a quantitative correlation between a Cepheid's average magnitude and its period of variability. Using such a correlation you will estimate the distance to a Cepheid in an external star system.

Introduction

Scientists are impressionable, and, just like other people, scientists have their heroes, too. The geologists have Wegener; the psychologists have Freud. The biologists have Darwin and Mendel, and the physicians have Harvey and Malpighi. The list of the heroes of physics and astronomy would be comprised of a long litany of names that certainly would include the likes of Newton and Einstein. These people were responsible for radical upheavals in their respective disciplines and for revolutions in thought, wholesale shifts in the way the discipline operated. In the works of these revolutionaries we may see the degree to which science is a collective and cumulative activity, and we may find in it the measure of the influence of an individual genius on the future of a cooperative scientific effort. In the work of these people we see that science advances primarily by heroic exercises of the imagination, rather than by the patient collecting and sorting of myriads of individual facts.

The mopping-up operation after a scientific revolution may be less glamorous and glitzy than being involved in the revolution itself, but important advances can be made by recognizing order and pattern in an assembly of data. Today's pattern may be tomorrow's law of physics.

In this experiment, you will correlate the average intrinsic brightness of a certain class of variable stars with the period of the brightness variations. If the intrinsic brightness (or, equivalently, the absolute magnitude) of any star is known, then the *distance* to that star can be estimated from the distance modulus relation

$$M - m = 5 - 5 \log_{10} d$$

where M is the absolute magnitude of a star, m its apparent magnitude, and d is the star's distance in parsecs (1 parsec = 3.26 light years). Without overstating the situation at all, the distance to an astronomical object is the single most important parameter needed for a full description of the object.

What do we mean by a correlation between the average intrinsic brightness of a star and its period of variability? In answer to this question, let us develop an analogy. Let us see if there is a correlation between height and weight in human beings. If we measured these quantities for a sizeable sample of people, we may find that the taller a person is, the more that person is likely to weigh. Such a correlation may be depicted graphically as in fig. 21.1.

The correlation between height and weight is not terribly precise; two people who weigh the same certainly need not have the same height. This is manifested in the graph by the vertical extent of the shaded region. The precision of the correlation can be improved somewhat, however. Human males and females develop differently, as you may have noticed, from 6 years of age to about 18, and peculiar things happen to both sexes around middle age. These are not large effects, at least in terms of the height-weight correlation, but the correlation could be improved somewhat by restricting the sample of people to a narrower age group, for instance males between the ages of 28 and 32. For this age group, the correlation would look like that depicted in fig. 21.2. Notice that the correlation between height and weight is tighter, more well-defined.

What this allows you to do is have some predictive power in the following sense: If you were told the weight of a person between the ages of 28 and 32, using the above graph (if it had numbers on it), you could tell within a few inches, what that person's height is. Or, conversely, if you knew a person's height, that person's weight would be known to within a few pounds. What sometimes happens is that in a well-intentioned but misguided effort to express the correlation more succinctly, a line (or some other curve) will be drawn through the data and the equation of the line (or curve) will be

Figure 21.1 Height-weight correlation for human beings of all ages.

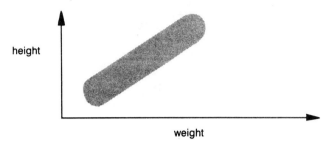

Figure 21.2 Height-weight correlation for human beings ages 28 to 32.

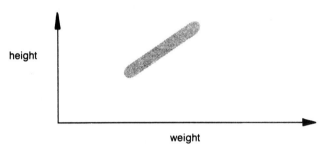

Figure 21.3 Linear fit to data in fig. 21.2.

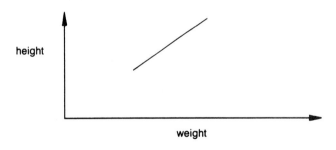

the only descriptor used for the correlation. In our analogy, the height-weight correlation would go to an intermediate step shown in fig. 21.3, and then the equation of the line would be determined.

For example, the correlation between height and weight might be expressed by the relation

$$H = 49.5 + 0.13W$$

where W is the weight in pounds, and H is the height in inches. Such a relation is useful because it portrays the general characteristics of the correlation between height and weight, namely that height increases linearly with weight. But the relation is potentially misleading on several counts:

1. One can get lulled into believing that there is something *fundamental* in the relation, that is, that it might be some sort of law. Clearly, this is not the case. The relation was derived from no general principle.
2. The relation masks any intrinsic variation in the data, that is for a given weight the relation yields only a single height for that weight. This of course does not reflect the fact that two people with the same weight can have different heights.
3. The character of the parent data is hidden. For example, it is not explicit in the relation that it is derived from data representing a limited sex and age group. The danger is that someone may use the relation outside its range of applicability.

Figure 21.4 Typical Cepheid light curve.

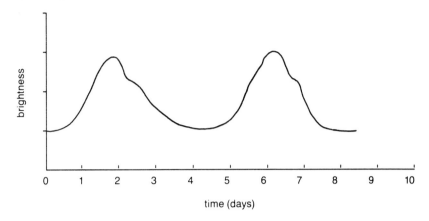

4. The relation gives nonsense when W = 0.

The moral of the story is that correlations between parameters can be quite useful if the limitations of the correlations are always kept firmly in mind. The advantage of discovering such an empirical correlation, apart from any practical utility it may have, is that it may guide the way to truly fundamental connections, the basic physical principles underlying the phenomenon.

Cepheid Variable Stars

Cepheid variables are giant stars with spectral types between F and K. The most rapidly varying ones complete a period—bright, faint, bright—in about two days, whereas the slowest take as long as 60 days. A plot of a variable's magnitude as a function of time, called its light curve, illustrates the periodic nature of the light variations (see fig. 21.4). Some Cepheids change their brightness by only 0.1 magnitude, whereas others vary by as much as 2 magnitudes.

Giant stars are rare, and giant stars that are Cepheids are even rarer. Nevertheless, some familiar stars are Cepheids. Polaris, for instance, is a Cepheid with a period of 3.9696 days and an amplitude of 0.1 magnitude.

Because Cepheids are intrinsically bright stars, they can be recognized not only in our own Galaxy, but can be seen in nearby galaxies (ones closer than about 13 million light years). A method of determining distances to Cepheids would be an important measuring technique, indeed.

The Experiment

Part 1 Period-Magnitude Relation for Cepheids in the Small Magellanic Cloud

To see if there is any correlation between intrinsic brightness and the period of variability of a Cepheid, we will need a sizable sample of Cepheids *all at the same distance*. If we had a sample of Cepheids at varying distances, then we would not be able to separate the effects of intrinsic brightness and distance on the apparent brightness of the Cepheid.

In order to find the correlation between intrinsic brightness and the period of variability, we will adopt the approach of Henrietta Leavitt, the discoverer of the correlation. Leavitt made a study of the Cepheids in the Small Magellanic Cloud (SMC), an irregular companion galaxy of our own Milky Way system. At the time, the distance to the Small Magellanic Cloud was not known with any precision, but it was clear that it was large compared to the size of the SMC itself. Consequently, the distance to the Cepheids in the SMC were all about the same, even though that distance was not known. Since all the stars in the SMC are at about the same distance, any correlation between period and apparent magnitude will also give a relationship between period and absolute magnitude.

Data and Analysis

We will be using Leavitt's data that appeared in her 1912 publication "Periods of 25 Variable Stars in the Small Magellanic Cloud." The data are presented in table 21.1. The data were obtained from a series of photographs.

In such a study, the detailed shape of the light curve is not as important as the times of maximum and minimum light. After monitoring several cycles, the period of a single cycle can be determined rather precisely, as table 21.1 shows.

Table 21.1. Periods of Variable Stars in the Small Magellanic Clouds

H.V.	Max.	Min.	Period	H.V.	Max	Min.	Period
1505	14^m.8	16^m.1	1^d.25336	1400	14^m.1	14^m.8	6^d.650
1436	14.8	16.4	1.6637	1355	14.0	14.8	7.483
1446	14.8	16.4	1.7620	1374	13.9	15.2	8.397
1506	15.1	16.3	1.87520	818	13.6	14.7	10.336
1413	14.7	15.6	2.17352	1610	13.4	14.6	11.645
1460	14.4	15.7	2.913	1365	13.8	14.8	12.417
1422	14.7	15.9	3.501	1351	13.4	14.4	13.08
842	14.6	16.1	4.2897	827	13.4	14.3	13.47
1425	14.3	15.3	4.547	822	13.0	14.6	16.75
1742	14.3	15.5	4.9866	823	12.2	14.1	31.94
1646	14.4	15.4	5.311	824	11.4	12.8	65.8
1649	14.3	15.2	5.323	821	11.2	12.1	127.0
1492	13.8	14.8	6.2926				

In table 21.1, the numbers in the H.V. column identify a particular star in the H(arvard) V(ariable) catalogue of stars. The apparent visual magnitude at maximum and minimum are given in the next two columns. The period of variability is listed in days.

As a first step in arriving at a brightness-period relation, plot the *average* apparent magnitude of the stars in table 21.1 vs. the logarithm of their respective periods. In order to make a large graph with reasonable resolution, you may want to omit the one or two stars with very long periods. Plotting the logarithm of the period rather than just the period yields a roughly linear relation, as you will see.

**What does the scatter of the data points about the line represent?

The plot you just made correlates apparent magnitude with period, but in order to use Cepheids as *distance* indicators, we need to know the distance to a Cepheid, any Cepheid.

Unfortunately, back in 1912 Leavitt was not able to determine the distance to any of the stars in the SMC. Nor could she find the distance to any Cepheids near the sun. Cepheids are rare, and it happens that not one is near enough to have a measurable parallax. (Polaris is just beyond the limit of current parallax capabilities.) Soon after Leavitt's work, however, Harlow Shapley used statistical methods to find the average absolute magnitude a few Cepheids near the sun.

P1. The result of such statistical studies (with some later refinements) is that a Cepheid with a 10.0 day period has an average *absolute* magnitude of -3.5. Use this information together with the distance modulus relation to determine the distance to the Small Magellanic Cloud.

P2. Next to the numbers on the apparent magnitude scale of your graph, write in the corresponding *absolute* magnitudes. This establishes a period-absolute magnitude relation.

Part 2 Distance to δ Cephei

δ Cephei is a star in our own Milky Way Galaxy and, as the name suggests, is the prototype of the class of variable stars we call Cepheids. The light curve of δ Cephei is given in fig. 21.5 below. As you can see from the apparent magnitude scale on fig. 21.5, δ Cephei is a fairly bright star, visible to the unaided eye. The variations in the brightness of δ Cephei were first reported by a deaf-mute astronomer, John Goodricke, in 1784.

Figure 21.5 Variations of apparent magnitude of δ Cephei.

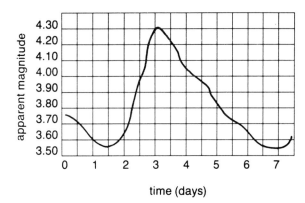

Figure 21.6 Equal area method for finding an average.

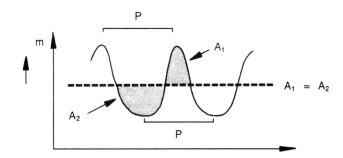

P3. Determine from fig. 21.5 the period P of the star δ Cephei. Calculate log P and from your graph of the absolute magnitude-period correlation, determine the absolute magnitude of δ Cephei.

In order to find the distance to δ Cephei, we still need to determine the average apparent magnitude of δ Cephei. One way would be to use the method utilized before, namely, the average apparent magnitude is the algebraic average of the extremes in magnitude, $(m_{max} + m_{min})/2$. An alternate way would be to find the magnitude that will give equal areas above and below a horizontal constant magnitude line (see fig. 21.6).

P4. Determine the average apparent magnitude of δ Cephei using the algebraic average method and the equal area method.

P5. Determine the distance D to δ Cephei using the average apparent magnitude you obtained with the algebraic average method. Get another estimate of the distance using the equal area method.

$D_{\text{algebraic average}} = \underline{\qquad}$ $D_{\text{equal area}} = \underline{\qquad}$

Distance to Cepheid Variable Stars

P6. You have two distance determinations using two different ways of estimating average magnitude. Is one distance measurement preferred over the other?

Part 3 Distance to an Unknown Cepheid

Use the data in the accompanying table 21.2 to determine the distance to the star whose light variations are represented. The magnitudes are photographic determinations similar to the ones used by Leavitt. The apparent magnitude of the star is given for the corresponding time of measurement in Julian dates. The Julian date, widely used in astronomy, is the number of the day in a running sequence beginning January 1, 4713 B.C., which in medieval times was thought to be the date of the creation of the earth. A Julian day begins at 12 noon one day and ends at noon on the next day, and so astronomical observations at night are all on the same Julian day.

Distance of unknown Cepheid = _____

P7. Is the star whose light curve is represented by the data in the Milky Way Galaxy?

Table 21.2 Variable Star Data

Date (Julian)	m_{app}	Date (Julian)	m_{app}
2438395.250	17.25	2438397.756	17.28
.288	17.30	.787	17.20
.376	17.32	.918	17.24
.500	17.35		
.614	17.38	2438398.000	17.30
.743	17.42	.166	17.31
.833	17.28	.254	17.28
.875	17.38	.333	17.30
.923	17.30	.417	17.32
		.503	17.36
2438396.011	17.22	.510	17.32
.042	17.00	.543	17.32
.084	16.50	.587	17.38
.166	15.99	.667	17.40
.239	15.82	.750	17.39
.261	15.84	.831	17.37
.376	16.03	.921	17.28
.496	16.20		
.521	16.21	2438399.000	17.20
.625	16.60	.035	17.05
.709	16.65	.063	16.48
.756	16.75	.166	15.90
.835	16.90	.239	15.82
.993	17.03	.261	15.80
2438397.081	17.09	.354	16.07
.166	17.10		
.250	17.00		
.292	17.05		
.500	17.19		
.564	17.21		
.625	17.30		

Exercise 22

Galactic Distance Determinations Using Novae

Learning Objectives

In this exercise, you will estimate the distance to a nearby galaxy by an analysis of the novae observed in the galaxy.

Introduction

In their theorizing and experimentation, physical scientists are guided by what can be characterized as a subliminal rule: The Principle of Minimum Astonishment. You will not find this explicitly described in your text, but it is there nonetheless. The Principle of Minimum Astonishment is the embodiment of an entire attitude, an outlook, that physical scientists have when dealing with the world and so cannot adequately be communicated in a few weak words. Regardless of its inadequacy, we will state the Principle of Minimum Astonishment in the following form: the least surprising answer is the correct one.

Implementing or applying the Principle of Minimum Astonishment necessitates that the person applying the Principle must have sufficient experience with the phenomenon in question in order to be surprised (or not surprised) by the answer to a question relating to the phenomenon. As an example, consider the result of releasing an object held in the hand, a stone or pencil. When asked whether the stone or pencil will move up or down or neither when released, the least astonishing result would be for the stone or pencil to fall down, and you respond accordingly. Your experience tells you that many objects when held in the hand and released fall down.

That the Principle of Minimum Astonishment is only a guide and not a rigid prescription is emphasized by what happens when you release from your hand a helium-filled balloon. Undoubtedly you are no longer surprised at the result, but perhaps at one point long ago you were.

One broad application of the Principle is particularly relevant to this laboratory session. Suppose you were asked the question, "Do the laws of nature (whatever they are) change from place to place and from time to time or are they the same, uniform everywhere and always?", how would you respond? It certainly seems that the rules, the laws, governing the operation of the cosmos do not get scrambled from place to place or from time to time. A stone released from my hand today where I am presently standing will fall just as it will tomorrow and in a different spot. There is no devious deity that is trying to confound our understanding of the world, surprising us at every turn. No, there is nothing special about our time and place we are told by the Principle of Minimum Astonishment. What we learn about the universe here and now is applicable tomorrow and everywhere.

In this lab experiment, we will make a study of novae in our own Milky Way Galaxy and compare their characteristics with those in a nearby galaxy. From such a comparison we will be able to deduce the distance to the galaxy. Of course, to make this comparison between novae in two different galaxies separated by vast stretches of distance and time we make the assumption, the unsurprising surmise, that the novae in the nearby galaxy behave as the novae in our Galaxy. Nothing guarantees that this is the case, but we believe with all our hearts that it is so.

The Nova Phenomenon

Stars are born out of immense clouds of gases and dust, and because a star once formed has a finite reservoir of fuel, they must ultimately die. Stars whose initial mass is less than about four solar masses (the limit is somewhat uncertain) end their lives as white dwarfs. More than 90% of the stars in our Galaxy are below this mass limit, so most stars will end their days as white dwarfs. A white dwarf is the exposed core of a once energetic star. With thermonuclear reactions having shut down, the core has collapsed to a density unheard of on earth—more than a ton per teaspoonful. With no energy-generating resources, the white dwarf simply cools slowly with the march of time.

A white dwarf need not be relegated to such an inauspicious demise if it is one component of a binary star system. Roughly half of the stars in the Milky Way Galaxy are members of a multi-component star system. For example, of the 33 nearest stars, 16 are in a multiple-star system. Two of those 33 stars are white dwarfs and each are members of a binary system, two stars orbiting each other about their common center of gravity.

Without overstating the matter at all, a star's mass is the single most important parameter that determines how the star will evolve. Two stars of roughly the same mass will evolve roughly in parallel. But a more massive star will spend its nuclear fuel faster, go through its evolution quicker, and be the first to become a white dwarf. Because binary systems are so common, there should therefore be, as there are, many cases of binary stars, one component of which is a white dwarf. Some such pairs are so close together that a thin trickle of the normal component's atmosphere flows onto the compact white dwarf. Having been robbed from the outer layers of the companion star, the material that gets deposited onto the surface of the white dwarf is hydrogen-rich. In time, the hydrogen accumulates, compressed to higher and higher pressures and temperatures by the intense gravity of the white dwarf, until the stolen atmosphere becomes so dense and hot that thermonuclear fusion reactions ignite on the white dwarf's surface. Only about 10^{-6}–10^{-4} solar masses of material need to accumulate for fusion reactions to be initiated.

As the thermonuclear fires rage across the surface, the detonation violently ejects the accumulated gases, creating an expanding shell of superheated material. Such an explosion increases the star's brightness by as much as 100,000 times. This sudden increase in the brightness of the star is called a **nova**, and to the ancient observers who witnessed such an event it seemed as though a new star had appeared in the heavens (*nova* is Latin for "new"). As the ejected shell grows larger, less dense, and cooler, the nova fades.

As spectacular and energetic as the nova phenomenon seems, the white dwarf and its companion are hardly affected. Mass transfer quickly resumes, again building a layer of hydrogen fuel and setting the stage for another blast in the future.

Part 1 Nova Cygni 1975

Perhaps 30–40 novae occur every year in our Milky Way Galaxy, but most are hidden from view by clouds of gas and dust that are found throughout the Galaxy. Occasionally, a nova becomes bright enough to be visible to the naked eye.

On August 29, 1975 a dramatic nova flared in the constellation Cygnus. It was observed by hundreds of amateur and professional astronomers around the world. Nova Cygni 1975 was somewhat atypical for a nova in two respects:

1. Most novae take two to four days to rise to maximum brightness and several months to decline. Nova Cygni 1975 rose to maximum just half a day after its discovery. The rapid rise was followed by an unusually rapid decline.
2. Neither the progenitor star nor any possible companion is visible on any photographs of the sky taken prior to the outburst. Researchers have deduced from studies of Nova Cygni 1975 that because the progenitor was so faint, it had to increase its luminosity at least 16 million times to reach the observed maximum brightness. Nova Cygni proved to be an unusually luminous nova.

There is some indication that Nova Cygni 1975 is a member of a close binary system with an orbital period of just over 3 hours.

Table 22.1 is a listing of the observed visual magnitudes of the nova at specified times. Plot the light curve of Nova Cygni 1975 on the coordinate grid provided.

Table 22.1 The Variation with Time of Nova Cygni 1975 (from Young, P. G., H. C. Corwin, Jr., J. Bryan, and G. de Vaucouleurs. *Astrophysical Journal*, 209: 882–894, published by the University of Chicago Press). (Copyright 1976 by the American Astronomical Society.)

No.	UT 1975	M_v
1	Aug 29.05	8.41
2	27.17	6.19
3	29.22	5.14
4	29.32	4.37
5	29.40	3.39
6	29.45	3.06
7	29.50	3.11
8	30.13	2.01
9	31.07	1.92
10	31.20	1.93
11	Sept 1.08	2.34
12	1.44	2.88
13	2.15	3.62
14	3.15	4.49
15	4.09	4.88
16	5.07	5.20
17	6.09	5.52
18	7.12	5.71
19	9.11	6.19
20	12.09	6.56
21	15.18	6.88
22	18.08	7.12
23	21.14	7.28
24	25.10	7.54
25	29.21	7.76
26	Oct 3.28	7.67
27	8.06	7.80
28	15.14	8.27
29	19.18	8.50
30	Nov 9.06	8.89

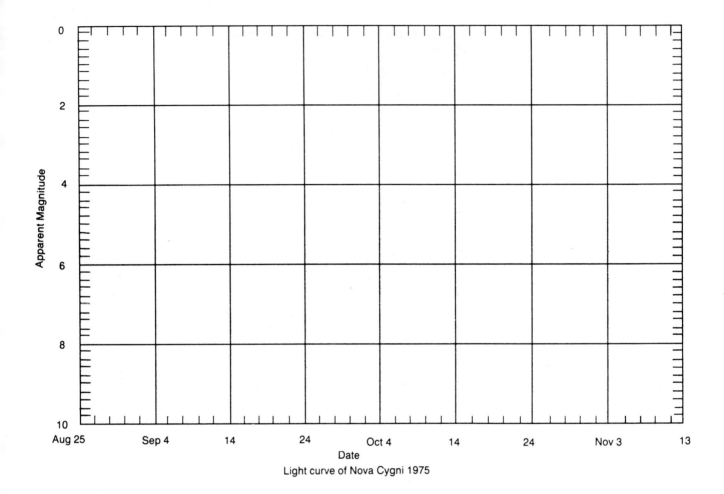

Light curve of Nova Cygni 1975

P1. Calculate the rate of rise of the nova in magnitudes per day.

Rate of rise = _____ mag/day

P2. When did the nova reach maximum luminosity? Be mindful of the level of accuracy with which you specify the answer.

The rate of decline of a nova is initially rather high; that is, a nova quickly dims after maximum light and this rate of decline varies from nova to nova. Nova Cygni, as mentioned previously, was an unusually fast nova, fast to rise and quick to decline. As a further complication, a nova's rate of decline is not constant. After a rapid dimming after maximum, a nova's decline becomes more and more gradual. We will characterize a nova's rate of decline by its *initial* rate of decline, which involves the time to fall 2 magnitudes from maximum.

$$\text{Rate of decline} = \frac{2 \text{ mag}}{\text{time to dim 2 mag from maximum}}$$

P3. Determine the time for Nova Cygni 1975 to fall 2 magnitudes from maximum.

Time to fall 2 mag from max = _____ days

P4. Calculate Nova Cygni's rate of decline in mag/day.

Rate of decline = _____ mag/day

Part 2 (Optional) Distance and Absolute Magnitude of a Nova

Under some circumstances, the distance to a nova can be estimated. Consider the case when we can see the shell of material that has been blasted out by the nova. If two photographs of the expanding shell are made several years apart, the rate of expansion can be deduced from the photographs. If, in addition, the actual velocity of expansion can be obtained from Doppler shift measurement, the nova's distance can be determined. Let us pursue this with a concrete problem.

Figure 22.1 Illustration of an expanding shell of gas ejected by a nova. At (a) the angular radius of the shell is 4.0 arc sec; at (b) 15 years later, the angular radius is 5.5 arc sec.

(a) (b)

Suppose two photographs of the expanding shell around a nova are taken 15 years apart; the photographs are represented schematically in fig. 22.1. In the first photograph, the radius of the shell is 4.0 arc sec, and in the second photograph taken 15 years later, the radius of the shell is 5.5 arc sec. Thus, the shell increased in size at the rate of (1.5 arc sec)/(15 years) = 0.1 arc sec/yr. The spectrum of the nova can reveal the velocity with which the ejected shell is expanding. Novae spectra are quite often complicated and confusing a few days or weeks after maximum, but become a bit easier to analyze later. Often in this later stage Doppler shift measurements become possible, and the velocity of expansion can be deduced. For our example, let us use a typical value of 1,000 km/sec. In 15 years, the shell would have expanded a distance

$$D = (\text{velocity})(\text{time})$$
$$= (1{,}000 \, \frac{\text{km}}{\text{sec}})(15 \text{ years})(3.16 \times 10^7 \, \frac{\text{sec}}{\text{year}})$$
$$= 4.74 \times 10^{11} \text{ km}$$

Figure 22.2 Determining the distance to a nova by an analysis of its expanding shell.

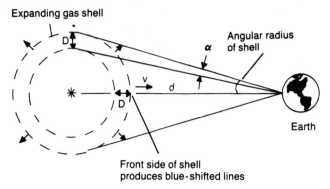

From the photographs, the shell grew by 1.5 arc sec (= 5.5 arc sec − 4.0 arc sec). This 1.5 arc sec corresponds to angle α in fig. 22.2 above.

P5. Use the trigonometric relation for thin triangles $\alpha = \dfrac{D}{d}$ to obtain the distance d to the nova. Note that 1 arc sec = 4.84×10^{-6} rad. (α in the above formula must be in radians). Express your answer in kilometers and parsecs. (1 parsec = 3.1×10^{13} km)

d = _____ km = _____ pc

P6. Suppose that the nova reached apparent magnitude m = +4.0 at maximum. Use the distance modulus relation $m - M = 5 \log_{10} d - 5$ to calculate the absolute magnitude M of the nova at maximum.

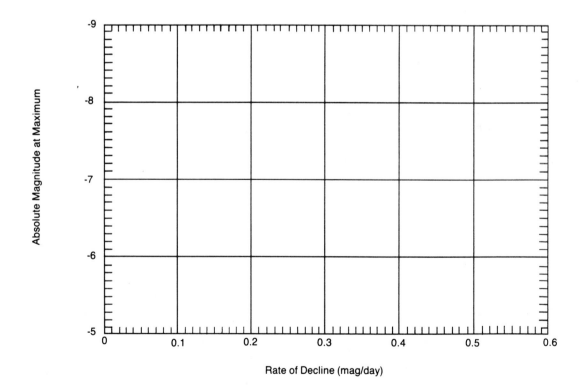

Data for novae in the Milky Way Galaxy

Part 3 Distance to the Andromeda Galaxy

Novae recorded throughout the years that occurred in our own Galaxy form an extensive body of data that is able to characterize the nova phenomenon. For the purposes of our experiment, it is seen (table 22.2) that the brighter the nova, the faster it declines. We will assume that the novae in the Milky Way Galaxy are fundamentally the same as those in the Andromeda galaxy, and that any correlation between absolute magnitude at maximum and rate of decline that we find in Milky Way novae will exist with Andromeda galaxy novae.

Table 22.2 below lists the absolute magnitude of novae that have occurred in the Milky Way Galaxy and their corresponding rate of decline. The rate of decline is determined by the time in days required for the novae to fade two magnitudes below maximum light, i.e.,

$$\text{Rate of decline} = \frac{2 \text{ mag}}{\text{number of days to fade 2 mag below maximum}}$$

Table 22.2 Characteristics of Novae in the Milky Way

Absolute Magnitude at Maximum	Rate of Decline (Magnitudes per Day)
−8.4	0.52
−8.35	0.57
−8.35	0.50
−8.3	0.40
−8.3	0.33
−8.3	0.30
−8.3	0.29
−8.25	0.25
−8.25	0.20
−8.25	0.17
−8.1	0.15
−7.95	0.14
−7.8	0.13
−7.7	0.12
−7.55	0.118
−7.35	0.105
−7.3	0.100
−7.15	0.091
−7.05	0.083
−6.75	0.064
−6.65	0.054
−6.6	0.050
−6.5	0.045
−6.45	0.040
−6.4	0.034
−6.4	0.032
−6.25	0.030
−6.2	0.027

For the novae in the Milky Way, plot their absolute magnitude at maximum versus the corresponding rate of decline on the coordinate grid provided.

Make a similar plot for the novae in the Andromeda galaxy, whose data are listed in table 22.3. Because the size of the Andromeda galaxy is small when compared to its distance from us, we can treat all the novae in the Andromeda galaxy as being at essentially the same distance.

To obtain the distance to the Andromeda galaxy, we will use the distance modulus relation

$$m - M = 5 \log_{10} d - 5,$$

where $m - M$ is called the distance modulus and the distance d is in parsecs (1 pc = 3.26 light years). We can directly obtain the distance modulus simply by superimposing the two graphs of peak magnitude vs. rate of decline. Place the graph of the Andromeda galaxy novae over the one for the Milky Way. Start with the horizontal axis of the two graphs being coincident, then slide the two graphs along the *vertical* axis until the distributions of novae on both graphs are superimposed. You may have to hold the graphs up to a bright light to see through the paper. Read any two corresponding values on the two magnitude axes to obtain $m - M$. Be careful when doing the subtraction to account for the negative absolute magnitudes of the Milky Way novae.

P7. Determine the distance modulus $m - M$ for the Andromeda galaxy novae.

$$m - M = \underline{\qquad}$$

P8. Calculate the distance to the Andromeda galaxy in parsecs, and convert to light years.

d = _____ pc

= _____ light years

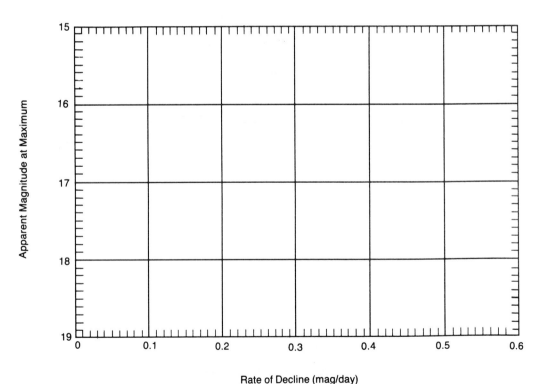

Rate of Decline (mag/day)

Data for novae in the Andromeda galaxy

Table 22.3 Characteristics of Novae in the Andromeda Galaxy (from H. C. Arp, *Astronomical Journal* (1956): 15–34. Copyright 1956 by the American Astronomical Society).

Apparent Magnitude at Maximum	Rate of Decline (mag/day)
15.9	0.38
18.2	0.20
15.9	0.17
16.0	0.23
15.9	0.15
16.0	0.23
16.0	0.174
16.0	0.29
16.1	0.180
17.0	0.077
16.2	0.163
16.4	0.126
16.7	0.157
17.2	0.069
17.5	0.058
17.6	0.070
17.2	0.060
17.4	0.075
17.6	0.067
17.4	0.046
17.8	0.059
17.6	0.061
18.0	0.043

Exercise 23

Globular Clusters: Distances and Ages

Learning Objectives

In this experiment, you will study the H-R diagrams of several globular clusters in order to estimate their distances and ages.

Introduction

Stars are born out of immense clouds of gases consisting primarily of hydrogen. Such clouds can contain many times—a million times—the mass of a single star. During the process of star formation, the cloud fragments into much smaller pieces, each collapsing separately from the others, each proceeding to give birth to a handful of stars.

These are the essential ideas in the making of stars, but the details of how stars are born remain elusive. The process of starbirth appears to follow two paths depending on the mass of the **protostar,** the collapsing mass of gas and dust out of which a star will be born when thermonuclear reactions begin in the interior. **Massive stars** seem to condense in small groups in bursts of star formation at the edges of the giant clouds. When they reach the main sequence as spectral class O and B stars, these massive stars with their enormous radiative energy output quickly and violently disrupt their mother cloud. In contrast, solar-mass stars appear to condense and form throughout the body of the giant gas clouds. When formed, they minimally affect their parent cloud. In all, the process of star formation does not appear to be very efficient. It locks up at most a small fraction of the cloud's mass as stable stars.

How a star lives and dies depends primarily, almost exclusively, on how much mass it has at birth. During its life, a star constantly struggles against gravity. Gravitational collapse is resisted by the pressure produced by the heat generated by thermonuclear fusion reactions in the core. More massive stars have higher core temperatures and produce energy at a higher rate than lower mass stars, i.e., their luminosities are greater. For example, a star with a mass 10 times that of the sun has about 2,000 times the sun's luminosity. *For main sequence stars,* this mass-luminosity relation is given by

$$\frac{L_{star}}{L_{sun}} = \left[\frac{M_{star}}{M_{sun}}\right]^{3.3} \tag{1}$$

The approximate masses for stars on the main sequence is indicated on the H-R diagram in fig. 23.1. Note that the most massive stars have the highest surface temperatures and also have the highest luminosities.

- P1. A star is twice as massive as the sun, i.e., $M_{star}/M_{sun} = 2$. Use the mass-luminosity relation given by eq. (1) to estimate how many times more luminous this star is compared to the sun. Check your answer by looking at fig. 23.1.

More massive stars may have more fuel available for fusion reactions, but these massive stars have higher core temperatures and thus higher luminosities. They produce energy at a furious rate, and even though they have more fuel than a lower mass star, they use up that fuel at a tremendous rate. Consequently, higher mass stars evolve much more rapidly than lower mass stars.

We can use the mass-luminosity relation to compare stellar lifetimes. The total amount of energy available to a star from the fusion of hydrogen to form helium is proportional to its mass. The rate at which the star produces energy is

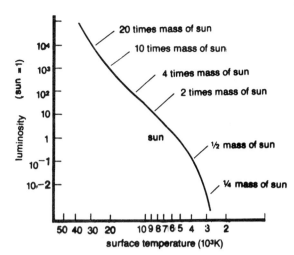

Figure 23.1 An H-R diagram with the main sequence indicated and approximate stellar masses at corresponding points on the main sequence.

given by its luminosity. Therefore, the time over which a star can radiate energy before its energy is used up is proportional to its mass divided by its luminosity, i.e., the energy available divided by the rate at which energy is used up. Relative to the sun, the lifetime of a star, t_{star}, with mass M_{star} and luminosity L_{star} is

$$\frac{t_{star}}{t_{sun}} = \frac{M_{star}/L_{star}}{M_{sun}/L_{sun}} = \frac{M_{star}}{M_{sun}} \frac{L_{sun}}{L_{star}}$$

If we substitute the mass-luminosity relation $L_{star}/L_{sun} = (M_{star}/M_{sun})^{3.3}$ in the above equation, we get

$$\frac{t_{star}}{t_{sun}} = \frac{M_{star}}{M_{sun}} \left[\frac{M_{sun}}{M_{star}}\right]^{3.3} = \left[\frac{M_{sun}}{M_{star}}\right]^{2.3} \qquad (2)$$

The star's lifetime, t_{star}, is *inversely* proportional to the 2.3 power of the star's mass. More massive stars have a shorter lifetime.

From models of stellar evolution, it has been determined that the main sequence lifetime of the sun is about 10 billion years (10^{10} yrs). In comparison, a 20 solar-mass star will have a lifetime of about

$$t_{star} = 10^{10} \text{ yr} \times \left[\frac{1}{20}\right]^{2.3} = 10^7 \text{ yrs} = 10 \text{ million yrs}$$

This star that has 20 times the mass of the sun goes through its evolution 1,000 times faster than the sun!

Globular clusters

Globular Clusters are aggregates of several hundred thousand stars gravitationally bound together forming a more or less spherical distribution. Figure 23.2 shows a typical example. It is a good bet that all the stars in a globular cluster formed approximately at the same time, at least any spread in age among individual stars is very small when compared to the age of the cluster itself. When such a cluster forms, all its stars have the same chemical composition, but have a range of masses. When we study an H-R diagram of a globular cluster, we will be directly investigating the effects of mass on the evolution of the stars. The more massive stars evolve more rapidly than the less massive ones. Thus, the more luminous stars evolve off the main sequence first, becoming red giants more quickly. As the cluster ages, stars of lower and lower mass will begin to evolve off the main sequence, that is, the point at which stars turn off the main sequence moves down to lower mass stars. Thus, the position of this *turnoff point* indicates the cluster's age. This is illustrated in fig. 23.3.

Figure 23.2 The globular cluster M13 in Hercules. Note how much more closely packed the stars are at the center of a globular cluster compared to its edges. In the center the stars are packed as high as 3 to 30 per cubic light year. If the sun were a star in the core of a globular, its nearest neighbors would be a few light months away. (Palomar Observatory Photograph)

Figure 23.3 The theoretical evolution of a cluster of stars born at the same time with the same composition, but having different masses. The dot indicates the position of a solar-mass star.

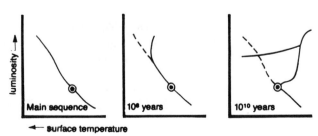

Figure 23.4 The thermal spectrum of a cool red star (left) and a hot blue star (right). The bandpasses for blue and red filters are indicated on each spectrum.

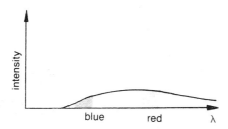
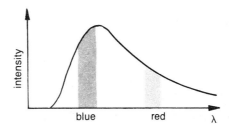

Color Magnitude Diagrams

The abscissa (the horizontal axis) on an H-R diagram is usually surface temperature or spectral class. (Historically, spectral class was used first.) Closely related to the surface temperature and spectral class is the idea of **color index.**

To see the connection between color index and temperature, consider differences in the spectra of a hot blue star and a cool red one. In our consideration, we will only worry about the continuum portions of the spectra and neglect any absorption lines (see fig. 23.4). The shape of the continuum is determined by the temperature of the star. Such a spectrum is also called a **thermal,** or **black body, spectrum.** The appearance of the thermal spectrum can be characterized by two quantities: the wavelength at which the spectrum reaches its maximum in intensity and the total power radiated, i.e., the star's luminosity. These two quantities depend on the temperature in the following way: the wavelength of the maximum is inversely proportional to the temperature,

$$\lambda_{max} \text{ (in meters)} = \frac{2.9 \times 10^{-3}}{T} \text{ (Wien's law)}$$

and the total energy radiated per second, the power output, is proportional to the fourth power of the temperature.

$$P \text{ (in watts per unit area)} = \sigma T^4 \text{ (Stefan-Boltzmann law)}$$
$$\text{where } \sigma = 5.67 \times 10^{-8} \text{ SI units}$$

Because the shape of the spectral curve is directly related to the temperature, we expect to be able to estimate a star's temperature if we can characterize its thermal spectrum.

Suppose we measure the brightness of a hot blue star through a blue filter and then make another brightness measurement through a red filter. A filter allows only a relatively narrow range of wavelengths to be transmitted; other wavelengths are absorbed. For example, a special blue filter we will be considering later allows light in a 900 Å band centered at 4,400 Å to pass through it. If we compare the brightness measurement of the star in the blue region of the spectrum and in the red region, the hot blue star is brighter in blue than in the red. The situation is reversed for the cool red star; the red bandpass brightness is greater than the blue bandpass brightness.

We use the language of magnitudes rather than brightness to express the relative amounts of blue and red light coming from a star. The **color index** of a star is the difference between its blue and red magnitudes, that is,

color index = blue magnitude − red magnitude
= B − R.

Actually, researchers use a variety of similarly defined color indexes; the variety originates because of different filters used. The color index we will use and is, in fact, most often measured is the B-V color index (for *B*lue-*V*isual). The blue filter is the one we described previously; the visual filter is an 800 Å bandpass filter centered on 5,500 Å in the yellow part of the spectrum (the filter is called 'visual' because our eyes are most sensitive to light in this region of the spectrum). A hot blue star has a higher blue brightness than a visual one; its blue magnitude is a *smaller* number than its visual magnitude, and its color index B-V is a negative number. A cooler star like the sun is brighter in the visual range than in the blue, and its color index is a positive number. (Remember that the magnitude scale runs "backwards"—the brighter the star, the smaller its magnitude.)

By means of the color index, we can determine the relative brightness of a star in two different wavelength regions and thereby characterize the shape of the thermal spectrum. This gives us the temperature of the star. Table 23.1 gives the correlations between temperature, spectral type, and B-V color index.

Table 23.1 Correlations between Surface Temperature, Spectral Type, and B-V Color Index. (From D. S. Hayes, "The absolute calibration of the H-R diagram," in *The H-R Diagram,* edited by A. G. D. Philip and D. S. Hayes, Reidel, Dordrecht, 1978.)

Temperature (K)	Spectral Type	Color Index (B-V)
47,000	O5	−0.32
30,300	B0	−0.29
15,300	B5	−0.16
9,410	A0	0.00
8,210	A5	+0.14
7,160	F0	+0.31
6,560	F5	+0.43
6,010	G0	+0.59
5,780	G5	+0.66
5,260	K0	+0.82
4,270	K5	+1.15
3,880	M0	+1.41
3,260	M5	+1.61

The Experiment

Part I Distances to Globular Clusters

The distance to a globular cluster can be estimated in a variety of ways. Often, a pulsating variable such as a RR Lyrae star or a Type II Cepheid can be distinguished among the myriad of stars in the globular cluster. For these kinds of stars, the period of variability is correlated to the star's absolute magnitude M. After the average apparent magnitude m of the star is measured, the distance modulus relation gives the distance d in parsecs (1 pc = 3.26 light years).

$$m - M = 5 \log_{10} d - 5 \tag{3}$$

The method we will use in this lab will be to determine the distance modulus $m - M$ by the process of main sequence fitting. In this method, we take a color-(apparent) magnitude diagram of a globular cluster and a color-(absolute) magnitude diagram of some nearby main sequence stars and shift the two diagrams along their vertical axes until the main sequences are superposed. To obtain the distance modulus $m - M$, all we need to do is compare any two corresponding points on the two vertical axes.

Figure 23.5 is a color-(absolute) magnitude diagram for a sample of nearby main sequence stars, and figs. 23.6–9 are color-(apparent) magnitude diagrams for four globular clusters. Remove these figures from the text so that the diagrams can be compared. Take the diagram for M13 and place it over the color magnitude diagram for the local main sequence stars. Hold the two diagrams up to a bright light or window and superpose the two color index scales. Now shift the two diagrams along their vertical axes until the main sequences on each diagram are aligned. Identify and record two corresponding points on the superposed vertical axes. That is, imagine a pin piercing the two diagrams at some convenient location on the superposed vertical axes—the vertical coordinates on each diagram and corresponding points—one is m and the other M. Record these numbers in table 23.2. Determine the distance modulus and the distance d in parsecs and in light years; enter your results in table 23.2.

Repeat the above procedure for the color-magnitude diagrams of M15 and NGC5466; M72 will be treated in a different way.

Table 23.2 Determination of Globular Cluster Distances

Object	Corresponding Points		Distance Modulus	d	
	m	M	m − M	Parsecs	Light Years
M13					
M15					
NGC5466					
M72					

Figure 23.5 Color magnitude diagram for a sample of nearby main sequence stars with well-determined absolute magnitudes.

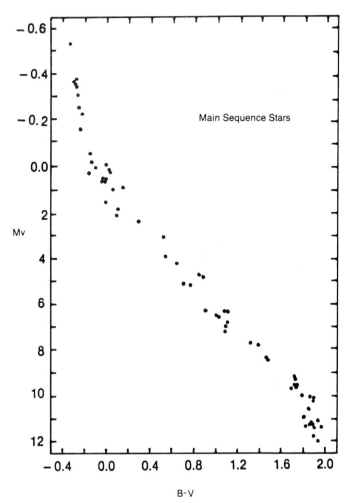

Globular Clusters: Distances and Ages

Figure 23.6 Apparent magnitude m_v vs. B-V color index for M13.

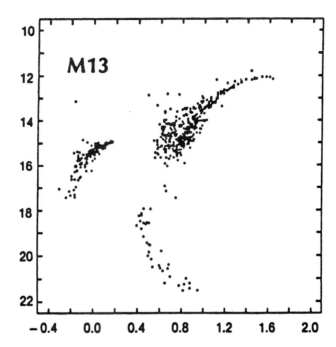

Figure 23.7 Apparent magnitude m_v vs. B-V color index for M15.

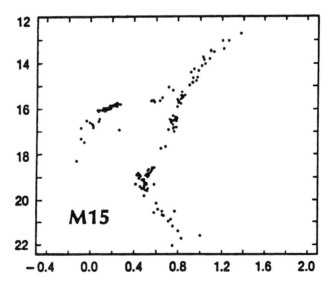

Figure 23.8 Apparent magnitude m_v vs. B-V color index for NGC5466.

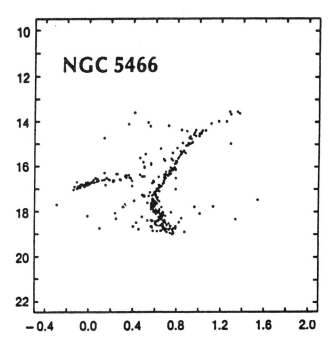

Figure 23.9 Apparent magnitude m_v vs. B-V color index for M72.

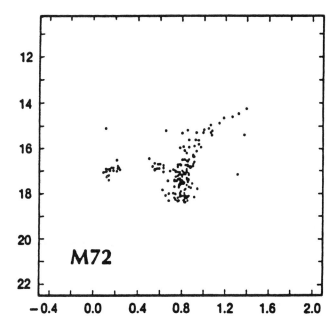

The survey that provided the data to construct M72's color-magnitude diagram only went as far as magnitude 18.5. The relatively faint main sequence stars do not appear. Since the main sequence is absent, we will be forced to make a separate comparison in order to measure the distance modulus of M72.

To obtain the distance modulus of M72, overlay its color-magnitude diagram with that of M13. This time superpose the giant branches of the two distributions. Note the difference between any two corresponding points. Adjust the distance modulus of M13 by this difference.

**Do you add or subtract this difference from the distance modulus of M13?

Calculate the distance d to M72 in parsecs and in light years, and enter the results in the previous table.

Part II Cluster Ages

From our previous discussion, the location of the main sequence turnoff point on a color-magnitude diagram is an indication of the cluster's age. A star at the turnoff point is just about to evolve off the main sequence, and so the age of a star at the turnoff point is the age of the entire cluster since all the stars were born very nearly at the same time. And we know how to calculate how long after a star is born it will leave the main sequence.

Equation (2) gives the main sequence lifetime of a star in terms of the sun's lifetime ($t_{sun} = 10^{10}$ yrs) and the ratio of the star's mass to the sun's. Our data do not contain the mass of a star at a turnoff point, but we have the next best thing, the apparent magnitude of a star at a turnoff point. The magnitude of a star is related to its luminosity, so we will have to change eq. (2) so that luminosities appear rather than masses. If the mass-luminosity relation in eq. (1) is substituted in eq. (2) to eliminate the masses, the result is

$$\frac{t_{star}}{t_{sun}} = \left[\frac{L_{sun}}{L_{star}}\right]^{0.7} \quad (4)$$

For M13, determine the apparent magnitude of a star at the turnoff point. From your distance modulus measurement of M13 in table 23.2, determine the absolute magnitude of a star at the turnoff point. Enter your results in table 23.3.

The absolute magnitude of a star is related to its luminosity by

$$\frac{L_{star}}{L_{sun}} = (2.512)^{M_{sun}-M_{star}} \quad (5)$$

where $M_{sun} = +4.8$, the absolute magnitude of the sun. (Astronomers the world over offer you humble apologies for the acursed magnitude scale.) Calculate L_{star}/L_{sun} for each turnoff-point star and use eq. (4) to calculate the age of the turnoff-point star and thus the age of the cluster. Note that the ratio L_{sun}/L_{star} appears in eq. (4).

P2. The age of the solar system is about 4.5 billion years. How does that compare with the globular cluster ages that you determined?

Table 23.3 Determination of Globular Cluster Ages

Object	m at Turnoff	M at Turnoff	$\frac{L_{star}}{L_{sun}}$	t_{star} (yrs)
M13				
M15				
NGC5466				

P3. Our Galaxy contains vast amounts of gas and dust. Most of the gas and dust is confined to the plane of the Galaxy and rapidly thins out above the plane. A distant object seen through this gas and dust has its light dimmed by passage through the intervening material. We did not take into account this dimming and so the apparent magnitude of a star at the turnoff-point is inaccurate. In the case of M15, for example, the dimming amounts to 0.4 magnitudes. Recalculate the age of M15, taking into account the dimming, and determine whether the presence of gas and dust causes you to overestimate or underestimate the ages of globular clusters.

Exercise 24: Hubble's Law

Learning Objectives

In this exercise, you will examine the correlation between the distance of a galaxy and its recessional velocity. This correlation, known as **Hubble's Law,** will be used to determine the distance to an unknown object using the object's spectrum.

Introduction

The universe began.

What an amazing deduction from a string of twentieth century observations! Ten or twenty billion years ago the universe, everything—matter, energy, *space,* and rules governing their interactions—originated in a cataclysmic explosion understatedly called the Big Bang. A few seconds after the Big Bang, the universe was concentrated at extremely high density. It was not that all the matter and energy were squeezed into a small corner of the present universe; rather, the entire universe, matter and energy, *and the space they fill,* occupied a very small volume. *You* are part of the Big Bang.

In that titanic explosion, the universe began an expansion that has never ceased. It is misleading to describe the expansion of the universe as a sort of distending bubble viewed from the outside. By definition, nothing we can ever know *was* outside. It is better to think of it from the inside, perhaps with grid lines imagined to represent the fabric of space expanding uniformly in all directions. As space stretched and expanded, the matter and energy in the universe expanded with it and rapidly cooled.

The early universe was filled with radiation and a plenum of matter, originally hydrogen and helium, formed from elementary particles in the dense primeval fireball. There was very little to see, if there had been anyone around to do the seeing. About a billion years after the Big Bang, the distribution of matter in the universe had become a little lumpy, perhaps because the Big Bang itself had not been perfectly uniform. A very small initial nonuniformity suffices to produce substantial condensations of matter later on.

Little pockets of gas, small nonuniformities, would steadily grow. Tendrils of vast, gossamer clouds formed—colonies of great, lumbering, slowly spinning things, steadily brightening—each a kind of beast eventually to contain a hundred billion shining points. The largest recognizable structures in the universe had formed. We see them today and we inhabit some lost corner of one. We call them galaxies. All the while the galaxies were all rushing away from each other, remembering their origins in the cataclysm that hurled their substance.

What are the characteristics of this expansion? Let us assume the expansion is uniform for the moment, and let us see if there should be a correlation between distance and recessional velocity. To make matters a bit simpler, let us look at a uniform expansion in one dimension (along a line) rather than the four dimensions of our universe.

Below is a string of galaxies, each separated by 1 unit of distance.

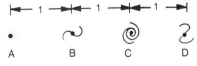

By "uniform expansion" we mean that a galaxy 1 unit of distance away moves away with 1 unit of velocity. As an example, let us say we are in galaxy B. Galaxies A and C are each 1 unit of distance away from B, so they each move away from B with 1 unit of velocity. For the sake of concreteness, let us say 1 unit of velocity is 150 km/sec.

153

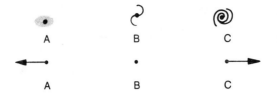

Galaxy D, however, is 1 unit of distance away from C, so it moves with a velocity of 150 km/sec away from C, but C is moving 150 km/sec relative to B. Therefore, relative to B, D is moving away with a velocity of 300 km/sec, or two units of velocity. D is two units of distance away and is moving away from B with two units of velocity—recessional velocity is, as you can see, directly proportional to distance in a uniform expansion:

$$V = Hd \qquad (1)$$

where H is the proportionality constant. In our universe, this number H has to be measured.

**Try the same analysis by picking A as the reference galaxy and showing that, relative to A, D is moving away at 450 km/sec = 3 units of velocity.

Notice that if you pick any galaxy as a reference, *all* other galaxies are moving away from it with a velocity that is directly proportional to distance.

Out in the "real" universe, if we measured the recessional velocities of galaxies using information in their spectra and somehow were able to determine their distances, the above relation in eq. (1) predicts a linear correlation between them; the slope of the line on a graph of V vs. d is the constant H, which is a measure of how fast the universe is expanding.

Using Albert Einstein's general theory of relativity, Alexander Friedmann in 1922 developed a model of the universe that was characterized by a uniform expansion as we have described. Experimental evidence for the expansion of the universe was furnished by Edwin P. Hubble in 1929. Below is the graphical display of the velocity-distance relation from Hubble's original 1929 paper. The value of the proportionality constant from his data is 513 km/sec per million parsecs; that is, for each million parsecs of distance, an object will be receding 513 km/sec.

Unfortunately, Hubble's determination of the constant was flawed for several reasons:

1. Hubble's determinations of the distances to the galaxies in his sample were inaccurate due to improper calibration of his distance indicators.
2. The galaxies in his sample nearer than 10^6 parsecs do not participate purely in the expansion of the universe, but their motions are affected by gravitational interactions among themselves.

Figure 24.1 Original diagram of the galactic velocity-distance relation from Hubble's paper "A relation between distance and radial velocity among extra-galactic nebulae." *Proceedings of the National Academy of Sciences 15* (1929):168–173.

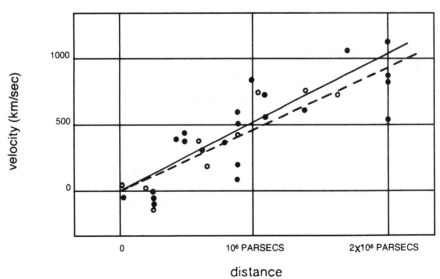

Figure 24.2 Images and spectra of elliptical galaxies near the centers of rich clusters. All the images are reproduced to the same scale. (Palomar Observatory Photograph)

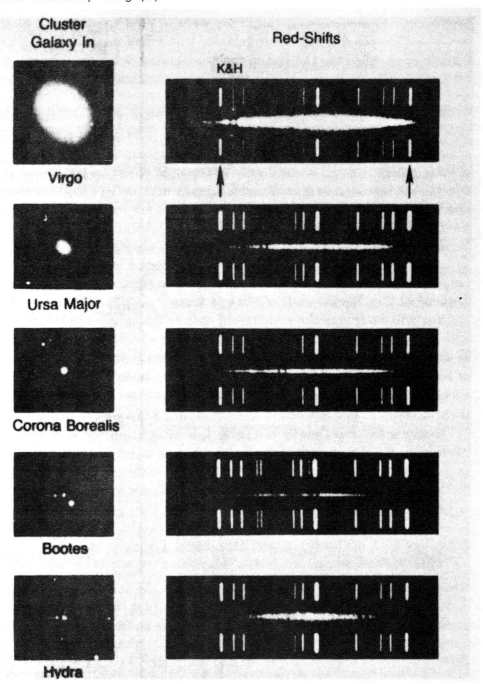

The Experiment

Figure 24.2 is a composite of five images of elliptical galaxies in different galaxy clusters and their respective spectra. The images of the galaxies are all reproduced to the same scale to facilitate comparisons among the images. Each of the elliptical galaxies in our sample lies near the center of a rich cluster of galaxies, a roughly spherically symmetric distribution of a thousand or so galaxies. Such an elliptical galaxy at the center of a cluster is frequently the most massive galaxy in the cluster. If we assume that all central elliptical galaxies at the centers of rich clusters are the same size, then the smaller the image of a galaxy, the further away it is. In fact, if one image is one-third the size of another, the object with the smaller image is three times further away. (This works fine for the small angular sizes of many things in the sky.) A more general statement is that a galaxy's distance is inversely proportional to its angular size *in our set of images*.

The spectrum associated with each galaxy appears on the right of the galaxy's image. Each spectrum is characterized by a nearly continuous streak running along the center of the panel. The emission lines of a comparison spectrum appear above and below the galaxy's spectrum; this comparison spectrum helps define the wavelength scale for the galaxy spectrum.

Two rather prominent absorption lines are visible breaking the continuity of the spectra of the galaxies. These closely spaced absorption lines are due to the presence of calcium and are called the **calcium H and K lines.** In laboratory spectra, these lines appear at wavelengths corresponding to 396.8 nm and 393.3 nm, respectively (1 nm = 10^{-9} m). But because the galaxies are receding, the spectra of these galaxies, including the calcium H and K features, are Doppler-shifted to longer (redder) wavelengths. As you can see from a quick examination of fig. 24.2, the greater the distance to a galaxy (the smaller the image size), the greater the red shift of the H and K lines.

The determination of the recessional velocities of the galaxies is made by measuring the shift in the wavelength of the H and K lines from their wavelengths in the laboratory. The laboratory wavelengths for the H and K lines were given above, but how do you measure the *observed* wavelengths on the spectra? The comparison emission spectra above and below the galaxy's spectrum calibrates the wavelength scale. Two arrows below the calibration spectrum of the Virgo galaxy point to two emission lines of helium. The left arrow points to a line of wavelength 388.8 nm, and the other arrow points to a line of wavelength 501.5 nm. So, all the points in between have wavelengths between 388.8 nm and 501.5 nm, a difference of 501.5 − 388.8 = 112.7 nm. Therefore, the observed wavelength of any feature in the spectrum of a galaxy a linear distance x to the right of 388.8 nm line can be computed from the relation

$$\lambda_{obs} = 388.8 \text{ nm} + \frac{x}{L} 112.7 \text{ nm} \tag{2}$$

where L is the linear distance between the 388.8 nm helium line and 501.5 nm helium line.

**Without simply plugging numbers into the above equation, determine the wavelength of a hypothetical feature in a galaxy's spectrum that is exactly halfway between the 388.8 nm line and the 501.5 nm line.

Figure 24.3 illustrates the anticipated effect of the motion of a galaxy on its spectrum.

It is easy to see from fig. 24.3 that the faster the source is moving, the more the wavefronts get piled up ahead of it and the more the wavefronts behind the source get drawn apart. The relation between the relative velocity v of the source and the shift in wavelength $\Delta\lambda = \lambda_{obs} - \lambda_0$ of a spectral line is

$$\frac{v}{c} = \frac{\Delta\lambda}{\lambda_0} = \frac{\lambda_{obs} - \lambda_0}{\lambda_0}, \tag{3}$$

where λ_0 is the wavelength of a spectral feature of a stationary source and c is the velocity of light (c = 3 × 10^5 km/sec). (The use of c for the speed of light is attributable to the Latin word *celer*, which means "swift," appropriate, you must agree.) The quantity $\Delta\lambda/\lambda_0$ is called the **red shift** of the object.

To determine a galaxy's recessional velocity, measure the distance L between the 388.8 nm and the 501.5 nm helium lines in the calibration spectrum. Next, measure the linear displacement x for the calcium H line (the "redder" one), and obtain λ_{obs} for this line for each galaxy using eq. (2). From eq. (3) you can calculate the recessional velocity of each galaxy (λ_0 for the calcium H line is 396.8 nm). Your results can be placed in table 24.1.

In order to correlate velocity of recession with distance, we will need, of course, to determine the distances to the galaxies in our sample. As was discussed previously, distance is inversely proportional to size for our similar galaxies. That is, a galaxy one-half the size of another is twice as far away. This gives only relative distances so far. To do better, we need to find the distance of one of the galaxies in our sample. The Virgo elliptical is near enough so that several methods can be used to estimate its distance. The result is that the Virgo galaxy is about 19 Mpc away. (Mpc = million parsecs = 10^6 pc; 1 pc = 3.26 light years.)

Measure the average size (in mm) of the galaxies in fig. 24.2. For the elliptical galaxies, a simple average of the long and short axes will be sufficient. For the smaller galaxies, try to measure to an accuracy of a few tenths of a millimeter. Using the inverse relationship of size and distance, and the known distance to the Virgo galaxy, determine the distances to the other galaxies. Enter your results in table 24.1. Note that the image of the central elliptical in the Hydra cluster is complicated by the fact that the image is of two closely spaced galaxies just barely resolved. Make your measurements on the leftmost galaxy image.

Figure 24.3 The Doppler effect. Successive wavefronts of a moving source S (successive positions marked by dots) occur closer to the observer ahead, decreasing the distance the wave must travel. Thus, the observer sees the wavefronts closer together relative to an observer behind the source who sees it moving away. The wavelengths are seen shortened ahead of the moving source (blue shift) and seen lengthened behind the source (red shift).

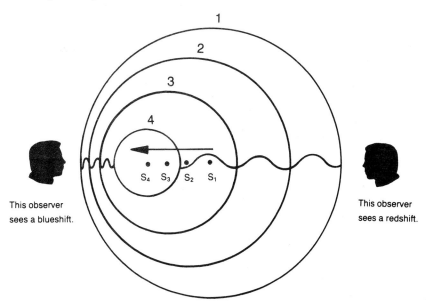

Table **24.1** Data for Hubble's Law					
Cluster	H Line Displacement x (in mm)	λ_{obs} (in nm)	Velocity (in km/sec)	Average Diameter of Galaxy (in mm)	Distance (in Mpc)
Virgo					19
Ursa Major					
Corona Borealis					
Bootes					
Hydra					

Make a full-page plot of velocity (vertical axis) against distance (horizontal axis) using your data from table 24.1.

**Besides the five galaxies in table 1, you have enough information to plot yet another point on your graph—the one for the Milky Way Galaxy. Where should you plot the Milky Way on the graph? Mark this point with an asterisk on the plot.

P1. Determine the value of the Hubble constant H from your graph [cf. eq. (1)]. The best approach is to draw a line on the graph such that it minimizes the scatter of the data points about the line. (You may find a clear plastic ruler useful for doing this.) Obtain a value for H by measuring the slope of the line. Be sure to include the units of H.

$$H = \underline{}$$

Figure 24.4 Spectrum of a galaxy.

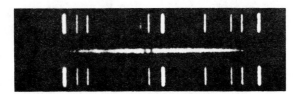

P2. From your calibration of the Hubble law, estimate the distance to the galaxy whose spectrum is given in fig. 24.4.

Discussion

The value of the Hubble constant has some rather interesting and important implications for cosmology. As we have seen, the Hubble constant is a measure of the rate of expansion of the universe. For a given distance away from an observer, the Hubble law tells how fast things are receding due to the expansion of the universe. The larger the Hubble constant, the faster the universe is expanding.

Question: Is the universe expanding so fast that the expansion will continue indefinitely, or is there a sufficient amount of matter in the universe with the mutual gravitational interactions among galaxies to eventually halt the expansion or even reverse it? Imagine trying to get something off the earth, for instance a stone. Hurling the stone upward with a velocity of a few meters per second just does not do it; the stone does not escape the earth's gravity. The stone goes up some maximal distance, stops, and returns. In order to keep the stone moving away indefinitely, you have to exceed some critical velocity for the stone, the escape velocity. For the earth, the critical velocity is 11.1 km/sec. So, too, the Hubble constant enters into answering questions about the eventual fate of the universe.

The Hubble constant is also related to the age of the universe. Suppose the galaxies have been moving with constant velocities since the Big Bang. Suppose also that the age of the universe is T. Then, if a particular galaxy had been moving with a velocity V during the time T, it would have moved a distance $d = VT$. But from the Hubble law, $d = V/H$ and so $T = 1/H$. From your measurement of the Hubble constant, you can estimate the age of the universe.

P3. From your value of H, calculate the age of the universe. Express your answer in billions of years and in seconds. Some useful conversions are:

$$1 \text{ Mpc} = 3.1 \times 10^{19} \text{ km}$$
$$1 \text{ yr} = 3.16 \times 10^{7} \text{ sec}$$
$$1 \text{ billion years} = 10^{9} \text{ years}$$

P4. Compare the age of the universe that you estimated with the age of the earth. Is it reasonable? (age of the earth = 4.5 ± 0.3 billion years)

The assumption of a constant velocity for a galaxy has an obvious complication: the gravitational interaction among galaxies acts to decelerate the expansion.

**If the expansion of the universe has been slowing since the Big Bang, would a particular galaxy have moved faster or slower in the past?

Another complication to the Hubble law originates in the finite velocity of light. When you look at a distant galaxy, not only are you looking deep into space, you are also looking deep into time. An object 1 billion light years away in space is 1 billion years back in time. Even though the linear Hubble law holds for any instant of time, the linearity is destroyed when looking over vast stretches of time: the Hubble "line" should curve slightly.

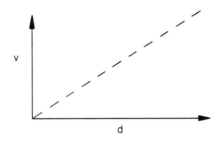

P5. Above is a Hubble law plot for an instant of time. For a decelerating expansion, determine if the Hubble velocity-distance correlation should curve upward or downward, and sketch it on the graph above.

P6. On the basis of the above, is the age of the universe that you determined larger or smaller than the actual age?

Exercise 25: Quasar Red Shifts and Distances

Learning Objectives

In this experiment, you will measure the red shift of the quasar 3C 273 from its spectrum and thereby deduce its distance. By measuring the spacing of the lines in the quasar's spectrum, you will also be able to check whether fundamental constants of nature have in fact remained constant over long stretches of time.

Introduction

Great surprises and the seeds of very great ideas often lie among the blizzard of entries in catalogues. In a compilation of the variable stars in the Magellanic Clouds, Henrietta Leavitt discovered a way to divine the distances to the galaxies. In a list of the red shifts of galaxies, Edwin Hubble showed us that beyond our own Galaxy the other galaxies recede in profusion in deepening space, the result of a universe expanding ever since its titanic inception billions of years ago. Catalogues may present a dry, uninspiring facade, but they can often lead to a hot commodity in science—understanding.

During the late 1950s, when radio astronomy was in its infancy, the locations of many radio sources in the sky were assiduously catalogued. It got to be something of a cottage industry for optical astronomers to find and photograph objects at the locations of radio emission. When these identifications were made, many of the objects turned out to be galaxies or nebulae! However, some of the objects appeared simply as stars—merely points of light—so the name **quasi-stellar radio sources** emerged, which was quickly shortened to **quasars**. Although the first quasars discovered were radio sources, most are not. When further study illuminated the peculiar aspects of the first quasars the term was broadened. The modern definition of a quasar refers to any very luminous starlike object with a large red shift. Herein lies the essence of the peculiar nature of quasars.

For a time they presented a unique puzzle to astronomers: the spectral lines from these objects simply did not seem to match any atoms known on earth. In 1963, Maarten Schmidt at the California Institute of Technology finally solved the problem. Schmidt obtained a spectrum of 3C 273, the 273rd entry in the third Cambridge catalogue of radio sources, and found a variety of lines that made no sense to him. After considerable reflection, he perceived the pattern. The light from this quasar, and others as well, had been shifted so far to the red that the spectrum had become unfamiliar and unrecognizable. It was as if someone had superimposed a map of New York City where we would normally expect to find one of Los Angeles. There is nothing particularly mysterious about the New York map, but until you figure out the shift you would certainly have trouble finding your way. In just the same way, spectral lines that astronomers expected to see in the visible region had been shifted into the infrared band in the quasar data, and the visible spectrum was populated by lines normally appearing in the ultraviolet range. Quite a disguise!

Schmidt discussed his ideas with a colleague down the hall and then spent the rest of the afternoon reassuring himself that he had overlooked nothing and was not somehow making a mistake. By nightfall, he and the colleague were satisfied that quasars did in fact display very large red shifts. The only way known for quasars to show such spectra, however, was for them to be at great distances in an expanding universe.

Schmidt drove home and told his wife, in his Dutch-accented English, "Something horrible happened at the office today." He saw her shocked expression and realized he should have chosen a different word. "I mean, something wonderful," he said.

Both words were appropriate. The lines in quasar spectra may have been identified, but if the red shifts were really as large as he claimed (and there was really no way out of that conclusion), and if we believe that large red shifts correspond to large distances as per Hubble's law, then quasars are among the most distant objects in the universe, billions of light years away. If quasars were really that far away, to be visible from earth they would need to produce enormous amounts of energy. From modern measurements we know that quasars typically disgorge radiation at rates 100–1,000 times greater than that of the Milky Way Galaxy. And with such active galaxies, we just do not have a good idea of the processes that produce this furious energy output.

In recent years a combination of computer-assisted search techniques and old-fashioned serendipity has turned up high red shift quasars in record numbers, high red shifts that correspond to distances in excess of 12 billion light years. The list of quasars at all red shifts now stands at about 5,000 and will continue to grow at an accelerating pace.

Since looking outward in space is equivalent to peering back in time, the conjecture is that quasars are nothing more than newly born galaxies. If by some magic you could see 3C 273 as it is today, it is possible that a galaxy much like the Milky Way would appear. Similarly, imaginary astronomers in 3C 273 looking at us might be seeing a quasar, a distant object in their own sky.

As best as we can tell, galaxies go through a series of evolutionary stages. A reasonable, although speculative, evolutionary scenario for a galaxy would have its start as a quasar, then move quickly through BL Lacertae, Seyfert, and similar active galaxy stages and wind up as a normal galaxy like our own. This is a very appealing idea. For one thing, it explains why the quasars are so far away; they may represent an early stage of galactic evolution, one close to the Big Bang. The scarcity of nearby quasars could suggest that the universe is no longer in the quasar-producing stage, and, by implication, no longer giving birth to galaxies. The strange and violent quasars we see in the universe may be like dinosaur skeletons, fossils of an earlier and harsher time.

The Experiment

Part 1 The Distance to 3C 273

The photograph in fig. 25.1 contains the spectrum of the quasar 3C 273 in which three hydrogen lines are identified. Under the spectrum of 3C 273 is a comparison spectrum that sets the wavelength scale for 3C 273's spectrum. The comparison spectrum also has three hydrogen lines identified as well as several other lines with corresponding wavelengths given in Angstroms (1 Å = 10^{-10} m). A quick look at 3C 273's spectrum shows that the hydrogen lines have all been shifted toward the red end of the spectrum. We assume that this shift is attributable to the Doppler effect arising from the recessional velocity of the quasar. The recessional velocity can then be computed from the (non-relativistic) Doppler formula

$$\text{red shift} \equiv z \equiv \frac{\lambda_{\text{observed}} - \lambda_o}{\lambda_o} = \frac{v}{c}, \tag{1}$$

where c is the speed of light (= 3×10^8 m/s) and λ_o is the wavelength as measured in the laboratory. This formula is adequate when the velocities are small relative to the speed of light. When v approaches c as in the case of many quasars, we must make use of the relativistic formula for the red shift z, namely

$$z = \frac{\sqrt{1 + v/c}}{\sqrt{1 - v/c}} - 1 \tag{2}$$

If we let $\lambda_{\text{observed}} - \lambda_{\lambda_o} = \Delta\lambda$ and solve eq. (2) for v/c, we get

$$\frac{v}{c} = \frac{\left[\frac{\Delta\lambda}{\lambda_o} + 1\right]^2 - 1}{\left[\frac{\Delta\lambda}{\lambda_o} + 1\right]^2 + 1} \tag{3}$$

Once we have obtained the recessional velocity v from the spectrum, we can use Hubble's law

$$v = Hd \tag{4}$$

to estimate the distance d to the quasar.

What you need to do now is determine the wavelengths of the hydrogen lines in the comparison spectrum and in the spectrum of 3C 273. Use the lines of known wavelength in the comparison spectrum to measure the wavelengths of the hydrogen lines in both the comparison spectrum and in the quasar spectrum. Enter your results in table 25.1.

Figure 25.1 The spectrum of the quasar 3C273 and a comparison spectrum used to get the wavelength scale. (Courtesy Maarten Schmidt, Palomar Observatory)

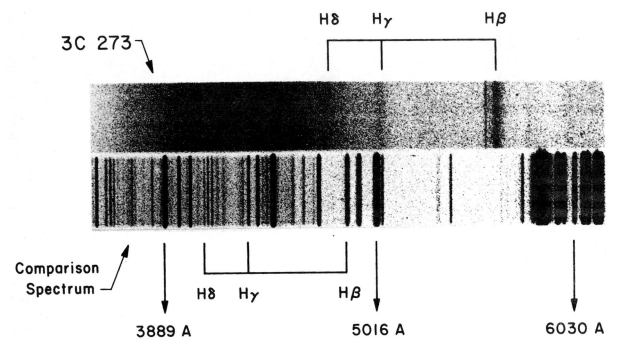

Table 25.1 Red Shift of the Quasar 3C 273			
Line	Laboratory Wavelength λ_o	Wavelength in Spectrum $\lambda_{observed}$	$z = \dfrac{\Delta\lambda}{\lambda_o}$
Hβ			
Hγ			
Hδ			

Calculate from your measurements $\Delta\lambda = \lambda_{observed} - \lambda_o$ and the red shift $z = \Delta\lambda/\lambda_o$ and enter these results in table 25.1.

Average your three values for the red shift $z = \Delta\lambda/\lambda_o$ and use the average value in eq. (3) to calculate v/c for 3C 273.

$$v/c = \underline{}$$

Hubble's law can now be used to estimate the distance to the quasar 3C 273. If you have done the Hubble's law lab prior to this one, use the value of it that you obtained in that lab. Otherwise, your instructor will provide you with a value.

$$d_{3C\ 273} = \underline{}$$

Optional

Equation (1) was described as the "non-relativistic" Doppler shift formula; eq. (3) was the relativistic analog of eq. (1). Let us investigate the differences between the two.

Using your average value for the red shift of 3C 273, recalculate v/c using eq. (1) rather than eq. (3).

$$v/c \text{ [from eq. (1)]} = \underline{}$$

Before we comment on this value, let us do the same calculations with a high red shift quasar. Surveys in recent years have turned up several quasars with red shifts larger than 4.0. Calculate v/c for a quasar with a red shift of 4.0 using both eqs. (1) and (3).

$$v/c \text{ [from eq. (1)]} = \underline{\hspace{2cm}}$$
$$v/c \text{ [from eq. (3)]} = \underline{\hspace{2cm}}$$

P1. 3C 273 is one of the nearest quasars. Compare the values of v/c calculated by the relativistic formula and the non-relativistic formula. How would the distance estimates from the two values of v/c compare?

P2. Answer question P1 in the case of the z = 4.0 quasar.

P3. For the z = 4.0 quasar, is any one of the v/c values "unacceptable"? Why?

Part 2 Stability of Fundamental Constants

The red shift of the hydrogen lines in the spectrum of 3C 273 depends upon the recessional velocity of the quasar. The *spacing* of the lines, however, also depends on some fundamental constants of nature; for example, the charge of the electron, the speed of light, and some other such constants. When you look at 3C 273 you are looking at ancient light, and so, by analyzing the spacing of the hydrogen lines in the spectrum of 3C 273 we can check on whether a certain combination of fundamental physical constants has changed since the time the light was emitted from the quasar.

For now, we will call the special combination of physical constants λ_R. Specifically,

$$\lambda_R = \frac{8\epsilon_o^2 h^3 c}{me^4},$$

where m is the mass of the electron, e its charge, h is Planck's constant, c the speed of light, and ϵ_o is the permittivity of free space. Quantum theory predicts that for the hydrogen lines we dealt with, wavelength differences would be

$$H\beta - H\gamma = \frac{12}{21}\lambda_R$$

$$H\beta - H\delta = \frac{5}{6}\lambda_R$$

$$H\gamma - H\delta = \frac{11}{42}\lambda_R$$

or solving for λ_R,

$$\lambda_R = \frac{21}{12}(H\beta - H\gamma) \tag{5a}$$

$$\lambda_R = \frac{6}{5}(H\beta - H\delta) \tag{5b}$$

$$\lambda_R = \frac{42}{11}(H\gamma - H\delta). \tag{5c}$$

Table 25.2 Determinations of λ_R at Two Different Times

	(1) λ_R Laboratory	(2) λ_R 3C 273	(1) − (2)
From $H\beta - H\gamma$			
From $H\beta - H\delta$			
From $H\gamma - H\delta$			

Equations (5a–c) are valid for the laboratory comparison spectrum. For the quasar spectrum

$$\lambda_R = \frac{1}{z+1} \frac{21}{12} (H\beta - H\gamma) \tag{6a}$$

$$\lambda_R = \frac{1}{z+1} \frac{6}{5} (H\beta - H\delta) \tag{6b}$$

$$\lambda_R = \frac{1}{z+1} \frac{42}{11} (H\gamma - H\delta) \tag{6c}$$

Notice that eqs. (6a–c) differ from eqs. (5a–c) only by a factor of $1/(z + 1)$, where z is the red shift, $z = \Delta\lambda/\lambda_0$. Equations (5a–c) allow us to calculate λ_R from the spacing of hydrogen lines in the laboratory comparison spectrum, and eqs. (6a–c) give λ_R from the spacing of the lines in the spectrum of 3C 273. Use your wavelength determinations from table 1 to calculate λ_R, and record your results in table 25.2.

The last column in table 25.2 contains three different determinations of the difference in λ_R from our time back to the time the light was emitted by 3C 273. To see if there is any significant difference in λ_R, average the results in the last column and calculate the mean absolute deviation of the three values.

The mean absolute deviation (MAD) was discussed in the experimental errors labs, but the following is a quick refresher. To calculate the MAD:

1. subtract each entry in the last column from the average value of the entries.
2. take the absolute value of each of those three differences from the average value.
3. average the three absolute values. This average is the MAD.

Express your results below in the form of average ± MAD.

$$(\lambda_{R,\text{ laboratory}} - \lambda_{R,\text{ 3C 273}})_{\text{average}} = \underline{} \pm \underline{}$$

P4. Has there been a change in the combination of physical constants represented by λ_R since the time of 3C 273?

Exercise 26: Entropy and Cosmology

Learning Objectives

In this experiment, you will be simulating a physical system with a fixed number of particles and constant total energy. By letting random exchanges of energy occur, you will examine the time evolution of the entropy of the system.

Introduction

Science is a collection of **models,** more or less quantitative descriptions of how the world works. The parameters used in the descriptions are human inventions that facilitate our thinking about the world. By now you have become familiar with many of the more important elements needed to describe the universe: concepts of energy, force, momentum, atoms, etc. It cannot be said that these things actually exist, whatever that means. There is no "energy," there are no "forces." The universe just *is*. We define some parameters that seem to be particularly useful, perhaps because they are conserved, and we build a model using the parameters.

Most scientists, once pinned down, would admit the models devised in science do not represent some absolute truth. What science *can* do is construct models that are *consistent* with observations and experiments. This is the principal measure of a scientific model: the model should be consistent with observations and experiments.

In this experiment, we will devise a useful parameter for physical systems, namely the **entropy** of the system. You will be simulating a physical system with a fixed number of particles and constant total energy. By setting up the system in a specific way (each particle with the same energy) and then letting random exchanges of energy occur, you will observe that the system approaches certain special distributions of energy that are, in the sense described below, "more likely" than others. A way to keep track of the approach of a physical system toward its more likely configurations is by examining the time evolution of the system's entropy. The entropy associated with any particular distribution turns out to be a measure of the "likeliness" or probability of the system achieving that distribution.

The Experiment

The apparatus for this lab consists of a board with 40 pegs (representing particles or atoms) arranged around a circle. A double spinner in the center is used to select pegs randomly. Washers placed on the pegs represent energy. An exchange of energy is simulated by moving a washer from one randomly selected peg to another.

Counting. To analyze the behavior of this system, we will begin by answering the question: How do we determine if one distribution of washers on pegs is more likely than another? The answer is that we must count the number of ways a distribution can occur. In physics terminology, we must count the number of **microstates** associated with a given **macrostate.** A simple example will illustrate the use of these terms.

For this example, our system will contain just three colored pegs (red, blue, green) and three identical washers. Establishing a "state" consists of putting the washers on the pegs so that every washer is on a peg. A typical state is illustrated below.

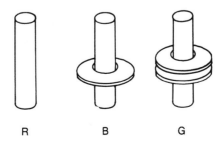

A complete description of a microstate for this example consists of specifying the number of washers on each colored peg. For the state shown, the microstate is described by "red has none, blue has one, green has two" or (R = zero, B = one, G = two).

In specifying a macrostate, one gives a general description of the system as a whole. The description consists of specifying *how many* pegs (regardless of color) are empty, how many hold one washer, how many hold two, etc., until all pegs are accounted for. It is as though we hid the colors of the pegs so that they appear identical. For the state shown, we would say "one peg has zero washers, one peg has one washer, and one peg has two washers," or more compactly (1 zero, 1 one, 1 two).

A little thought reveals that there are three possible macrostates for our simple system: (2 zeroes, 1 three), (1 zero, 1 one, 1 two), and (3 ones). How do we count the number of microstates associated with each? One way is to list them exhaustively: try this for (2 zeroes, 1 three). You should find 3: (R = three, B = zero, G = zero), (R = zero, B = three, G = zero), (R = zero, B = zero, G = three). The methods of counting combinations and permutations like these give the result that

$$W = \frac{N!}{N_0! N_1! N_2! \ldots} \tag{1}$$

where W = number of microstates associated with the macrostate (N_0 zeroes, N_1 ones, . . .), N = total number of pegs (particles), and N_n = number of pegs holding n washers (energy units). Also, recall that zero factorial is defined as $0! = 1$.

**Verify using eq. (1) that there are three microstates associated with the macrostates (2 pegs with zero washers and 1 peg with three).

Note the following rules for this system:
Rule 1: $N = N_0 + N_1 + \ldots$ Conservation of pegs (number of atoms)
Rule 2: $E = 0 \cdot N_0 + 1 \cdot N_1 + 2 \cdot N_2 + \ldots$ Conservation of washers (energy)

P1. Calculate 6!

P2. Calculate W for each macrostate described in the 3-peg system above. Check your results by an exhaustive list. Which is the "most likely" distribution?

Entropy. Astrophysicists often observe large systems of atoms continuously exchanging energy through collisions. We describe the systems **macroscopically**—typically giving temperature, pressure, etc. We could never hope to list the energies of the $\sim 10^{23}$ molecules in a typical system. We assume that all possible microstates are equally likely and that the system will therefore evolve toward those macrostates associated with the maximum number of microstates. As an example, consider a deck of playing cards, newly purchased with all the cards in "order": spades, A,2,3, . . . ; diamonds, A,2,3, . . . etc. (the order is somewhat dependent on the game played and on other imponderables). Start shuffling the deck. Soon the initial order is lost. If you continue shuffling, continue random reorganization of the cards' positions, would you want to wait around for the initial order to reemerge? Of course not! Out of all the possible combinations of 52 cards, the initial configuration is only *one*. The number of possible configurations in which the deck looks "shuffled" is so huge that in randomly selecting different deck configurations, you would have to wait an unpleasantly long time until the initial ordered state happened to appear.

Thus, when we observe a system evolving, we might expect W [from eq. (1)] to increase. It turns out that the logarithm of W is more convenient. In fact, one way to define entropy S is

$$S = k \ln(W) \tag{2}$$

where k is a constant.

Procedure

1. You will model a system of 40 atoms with an average energy per atom of either one unit or two units (check with your lab instructor for which case to do). Begin by placing either one washer on each post or two washers on each post (depending on which case you are doing).

    An energy exchange consists of transferring a single washer (unit of energy) from one post (atom) to another. Spin both spinners. Remove a washer from the post indicated by the long spinner and place it on the post indicated by the short spinner. If the long spinner points to an empty post, spin both again. (Respins *do not count* as an exchange.)

2. Now perform 100 exchanges, pausing frequently to record the distribution and calculate the entropy. Record the distribution after 5 exchanges, then after 10, 20, 30, . . . 100 (every 10).

Make a chart in your notebook like the sample below.

Number of Posts with n Washers after ℓ Exchanges					
n\ℓ	0	5	10	20	. . . 100
0	—		9		19
1	40		24		10
2	—		5		6
3	—		2		3
4	—		—		1
5	—		—		1
6	—		—		—
7	—		—		—
8	—		—		—
9	—		—		—
S	0		37.26		47.51

The chart shows how to record a distribution. Initially there were no empty posts, 40 posts with one washer, no posts with two or more. After 10 exchanges there were 9 empty posts, 24 with one, 5 with two, and 2 with three.

Here is how to calculate entropy, using eqs. (1) and (2) and the table at the end of these instructions. Initially we have $N = 40$, $N_0 = 0$, $N_1 = 40$, $N_2 = 0, \ldots$. Thus, eq. (1) gives

$$W = \frac{N!}{N_0! N_1! N_2! \ldots} = \frac{40!}{0! 40! 0! \ldots} = \frac{40!}{40!} = 1.$$

Since $0! = 1$, blank entries in your distribution chart may always be ignored.

To calculate entropy for the initial distribution, choose units in which $k = 1$ in eq. (2). Then

$$S = k \ln W = (1) \ln (1) = 0$$

and record this in the chart.

For most distributions W is incredibly large. Thus, proceed as follows. Write an expression for W. For example, after 10 exchanges we find the distribution from the chart above and write

$$W = \frac{40!}{9! 24! 5! 2!}$$

Then use the properties of logarithms and table 26.1. We find

$$\begin{aligned} S &= \ln W \\ &= \ln 40! - (\ln 9! + \ln 24! + \ln 5! + \ln 2!) \\ &= 110.32 - (12.80 + 54.78 + 4.79 + 0.69) \\ &= 37.26 \end{aligned}$$

3. Plot S vs. t (number of energy exchanges). Use a full page for the plot.
4. Find a lab group that did the other case (i.e., the case you did not do), and copy and study the two plots from step 3 that this lab group obtained. You should copy these plots on the same axes as you used for your plots. Treat the data copied as your own; i.e., treat the other lab group and your group as one large lab group.

Entropy and Cosmology

Table 26.1 Integers and the Natural Logarithms of Their Factorials

n	ln(n!)	n	ln(n!)
40	110.32	20	42.34
39	106.63	19	39.34
38	102.97	18	36.40
37	99.33	17	33.51
36	95.72	16	30.67
35	92.14	15	27.90
34	88.58	14	25.19
33	85.05	13	22.55
32	81.56	12	19.99
31	78.09	11	17.50
30	74.66	10	15.10
29	71.26	9	12.80
28	67.89	8	10.60
27	64.56	7	8.53
26	61.26	6	6.58
25	58.00	5	4.79
24	54.78	4	3.18
23	51.61	3	1.79
22	48.47	2	0.69
21	45.38	1	0.00
		0	0.00

P3. For each case, does your system move uniformly toward more and more likely distributions? Why or why not?

P4. What happens to entropy as more energy becomes available to a system with a fixed number of atoms?

P5. Two systems were examined in this experiment: 1) $E_{average}$ = 1 energy unit per atom, and 2) $E_{average}$ = 2 energy units per atom. Which system is "hotter"?

P6. What would your S vs. t plot look like if $E_{average}$ was 3 energy units per atom? A small sketch might be the most concise answer.

P7. How does the entropy of a system change with time?

The connection of entropy with time is very interesting (and complicated). As illustrations, three "movies" will be described, none of which would be a contender for the Cannes Film Festival award.

1. A movie of a closeup of a pool table, which shows two moving pool balls colliding.

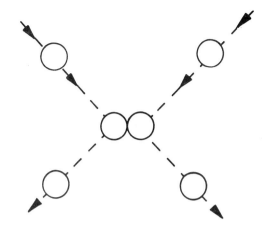

Is it possible to tell whether the movie is being run forward or in reverse? The answer is "no."

2. A movie of the break in a pool game.

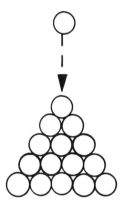

Is it possible to tell whether this movie is being run forward or in reverse?

P8. How do *you* know the answer is "yes"?

3. A movie of an ice cube. Run one way we see the ice cube melt. Run the other way we see a puddle of water collect and freeze to form an ice cube. Can you tell the direction of time here?

As peculiar a concept as entropy might seem, human beings are good at spotting changes in entropy.

Your observations concerning the time evolution of the entropy of an isolated system constitute the Second Law of Thermodynamics: entropy either remains constant or increases in an isolated system. An example of such a system is the universe itself, and the concept of entropy is quite useful in describing the time evolution of the cosmos.

As you saw from the experiment, the entropy of a system depends on the total number of particles in the system and the number of interactions between the particles. The greater the number of particles or number of interactions, the greater the entropy; that is, the entropy of a system is proportional to the number of particles in the system.

What are the most abundant particles in the universe? In the universe in its present epoch, photons and neutrinos are the most abundant particles. The number of photons per cubic meter of thermal radiation at temperature T is

$$n_{photons} = 2 \times 10^7 \, T^3$$

and for the universe now at a temperature of about 2.7K the number of photons per cubic meter is about 400 million. The number of neutrinos is approximately the same as the number of photons.

Let us compare the entropy of matter (essentially the total number of nuclear particles out of which the galaxies are made) with the entropy of the photons (which is proportional to the total number of photons). The number density of nuclear particles—protons and neutrons—in the universe today is about one nuclear particle per cubic meter. (There is considerable uncertainty in this number. It could be ten times as large, but that uncertainty will not affect our description significantly.) The number of photons per cubic meter is about 400 million, a number determined by its current temperature of 2.7K. Hence, the ratio of the entropy in photons to that of nuclear matter, the **specific entropy,** is 400 million (with an uncertainty of about a factor of 10). The neutrino-specific entropy is about the same. We see that the entropy of the universe today is almost all in the background gas of photons and neutrinos that fills the universe, not in nuclear matter.

The universe is a closed system, and hence its entropy, which we see is mostly in the gas of photons and neutrinos, increases in time in accord with the Second Law of Thermodynamics. Galaxies form and stars burn, thus dumping more photons and neutrinos into space and adding to the photons and neutrinos that have existed since the beginning of the universe. These processes increase the total entropy of the universe. But the remarkable fact is that the increase in the total entropy of the universe, from all these processes integrated over the entire lifetime of all the galaxies and stars, is only *one ten thousandth* of the entropy already in the background photons and neutrinos—a tiny fraction. For all intents and purposes, the entire entropy of the universe today is in the photons and neutrinos that pervade the cosmos and has remained effectively constant since the Big Bang. Entropy is essentially a conserved quantity in our universe.

Not so long ago scientists spoke of the "heat death" of the universe. In the 1930s the physicist James Jeans, reflecting the view of most of his colleagues, remarked:

> For, independently of all astronomical considerations, the general physical principle known as the second law of thermodynamics predicts that there can be but one end to the universe—a "heat death" in which the total energy of the universe is uniformly distributed, and all the substance of the universe is at the same temperature.

Physicists like Jeans who realized the universe was subject to the Second Law of Thermodynamics were not wrong about the heat death. But they did not know in the 1930s about the existence of the photon gas at 2.7K and the neutrino gas at about 2K. We now know that the "heat death" of the universe *happened long ago* with the Big Bang that created those particles. All the stars burning out can contribute but a tiny fraction to the total entropy that is already there. We live in a universe that has very nearly reached its maximum entropy.

Appendix 1 Astrophysical

Fundamental Constants

Speed of light	$c = 2.9979 \times 10^8$ m/s
Planck's constant	$h = 6.6262 \times 10^{-34}$ J·s
Gravitation constant	$G = 6.672 \times 10^{-11}$ N·m²/kg²
Unit charge (electron and proton)	$e = 1.6021 \times 10^{-19}$ C
Electron mass	$m_e = 9.109 \times 10^{-31}$ kg
Proton mass	$m_p = 1.673 \times 10^{-27}$ kg
Neutron mass	$m_n = 1.675 \times 10^{-27}$ kg
Boltzmann's constant	$k_B = 1.38 \times 10^{-23}$ J/K
Stefan-Boltzmann constant	$\sigma = 5.67 \times 10^{-8}$ W/m²K

Astronomical Constants

Earth: mass = 5.98×10^{24} kg
 radius = 6.38×10^6 m
Moon: mass = 7.35×10^{22} kg
 radius = 1.74×10^6 m
Sun: mass = 1.99×10^{30} kg
 radius = 6.96×10^8 m

astronomical unit (AU) = 1.496×10^{11} m
earth–moon distance = 3.84×10^8 m

Luminosity of the sun = 3.90×10^{26} W
Solar constant = 1370 W/m²

1 parsec = 206265 AU = 3.26 ly = 3.086×10^{16} m
1 light year (ly) = 9.46×10^{15} m = 6.32×10^4 AU

Metric Prefixes

milli	$m = 10^{-3}$	kilo	$K = 10^3$
micro	$\mu = 10^{-6}$	mega	$M = 10^6$
nano	$n = 10^{-9}$	giga	$G = 10^9$
pico	$p = 10^{-12}$	tera	$T = 10^{12}$
femto	$f = 10^{-15}$	peta	$P = 10^{15}$

Appendix 2 Maps of the Evening Sky

THE NIGHT SKY IN JANUARY
(*Griffith Observer* monthly magazine, Griffith Observatory, Los Angeles)

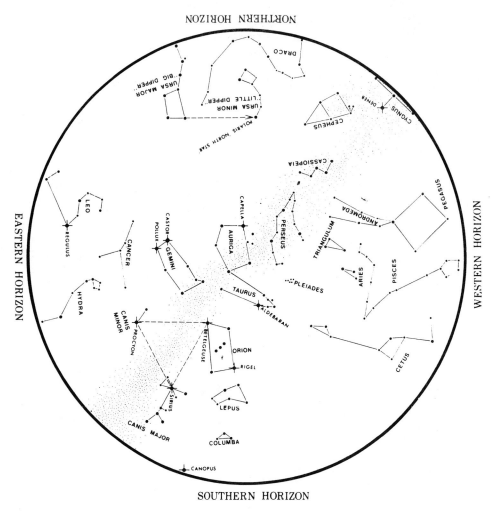

THE NIGHT SKY IN JANUARY

174 Appendix 2

THE NIGHT SKY IN FEBRUARY
(*Griffith Observer* monthly magazine, Griffith Observatory, Los Angeles)

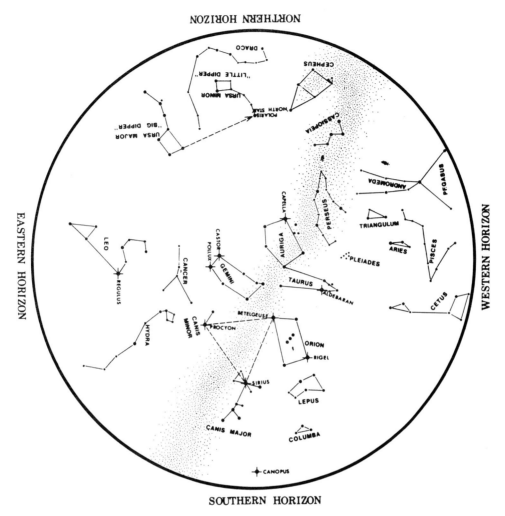

THE NIGHT SKY IN FEBRUARY

Maps of the Evening Sky

THE NIGHT SKY IN MARCH
(*Griffith Observer* monthly magazine, Griffith Observatory, Los Angeles)

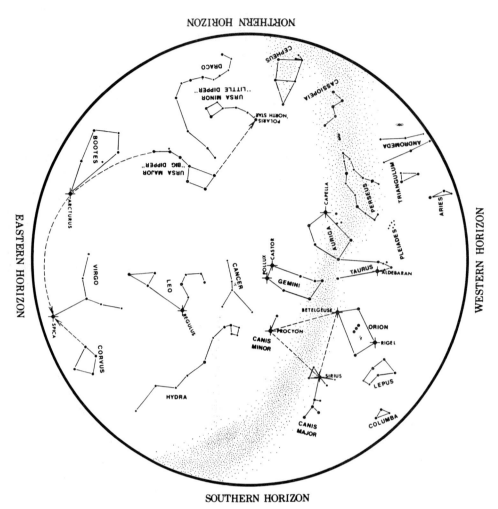

THE NIGHT SKY IN MARCH

Appendix 2

THE NIGHT SKY IN APRIL
(*Griffith Observer* monthly magazine, Griffith Observatory, Los Angeles)

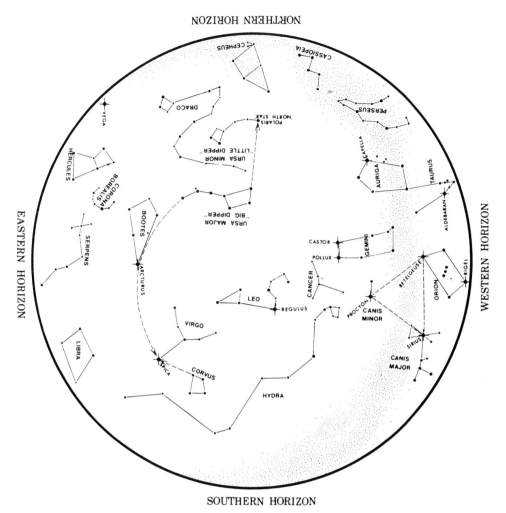

THE NIGHT SKY IN APRIL

THE NIGHT SKY IN MAY
(*Griffith Observer* monthly magazine, Griffith Observatory, Los Angeles)

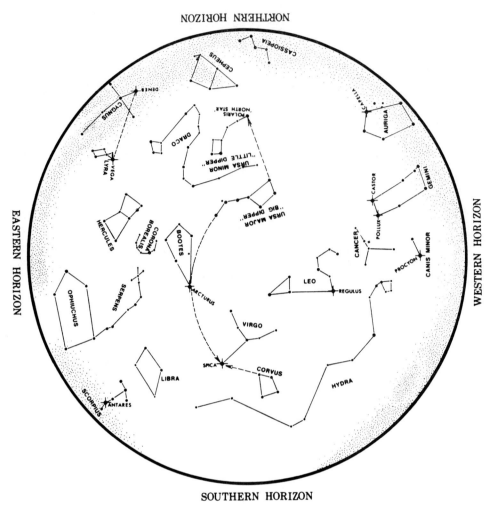

THE NIGHT SKY IN MAY

178   Appendix 2

THE NIGHT SKY IN JUNE
(*Griffith Observer* monthly magazine, Griffith Observatory, Los Angeles)

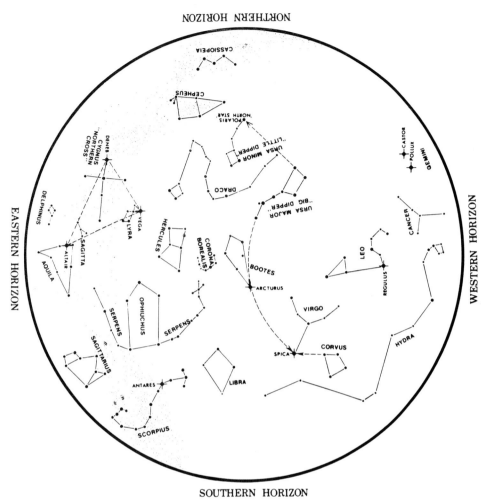

THE NIGHT SKY IN JUNE

THE NIGHT SKY IN JULY
(*Griffith Observer* monthly magazine, Griffith Observatory, Los Angeles)

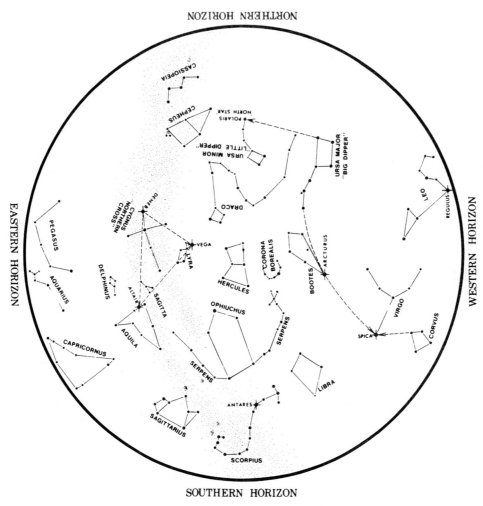

THE NIGHT SKY IN JULY

180   Appendix 2

THE NIGHT SKY IN AUGUST
(*Griffith Observer* monthly magazine, Griffith Observatory, Los Angeles)

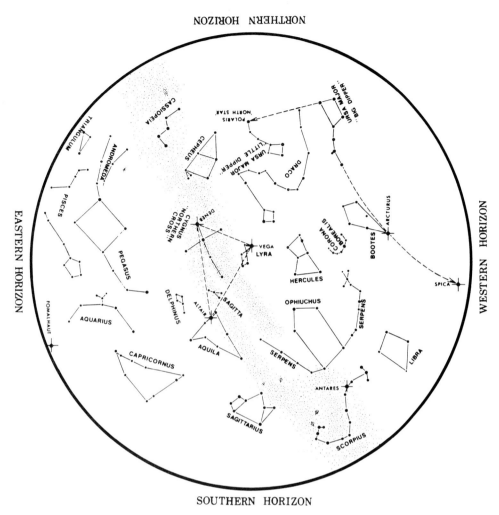

THE NIGHT SKY IN AUGUST

Maps of the Evening Sky

THE NIGHT SKY IN SEPTEMBER
(*Griffith Observer* monthly magazine, Griffith Observatory, Los Angeles)

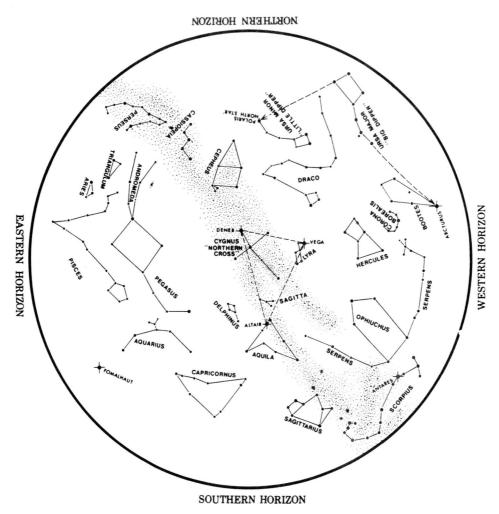

THE NIGHT SKY IN SEPTEMBER

THE NIGHT SKY IN OCTOBER
(*Griffith Observer* monthly magazine, Griffith Observatory, Los Angeles)

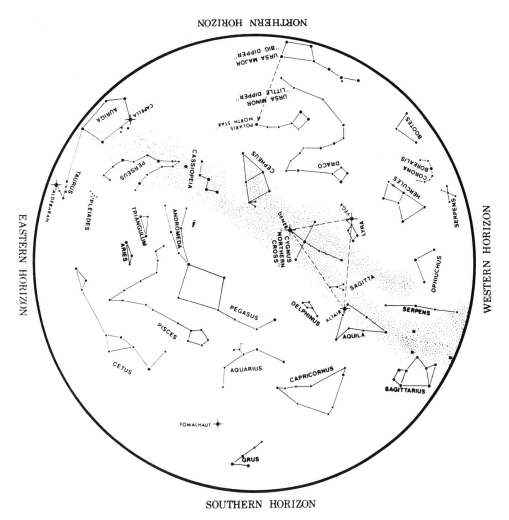

THE NIGHT SKY IN OCTOBER

Maps of the Evening Sky 183

THE NIGHT SKY IN NOVEMBER
(*Griffith Observer* monthly magazine, Griffith Observatory, Los Angeles)

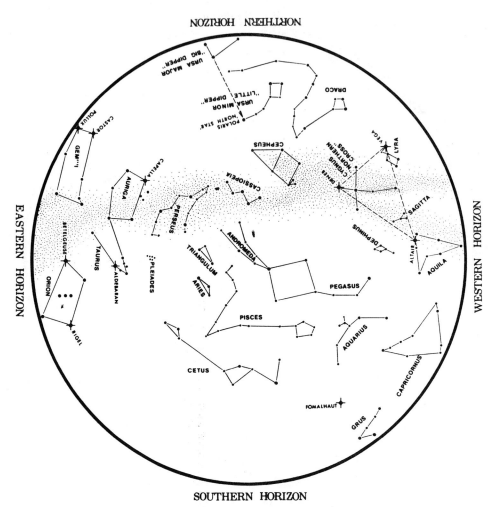

THE NIGHT SKY IN NOVEMBER

184   Appendix 2

THE NIGHT SKY IN DECEMBER
(*Griffith Observer* monthly magazine, Griffith Observatory, Los Angeles)

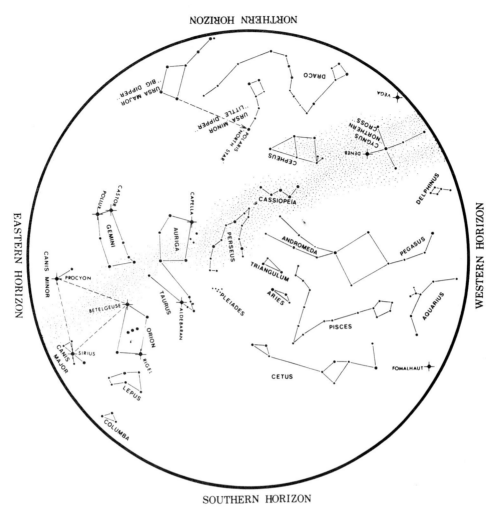

THE NIGHT SKY IN DECEMBER

Maps of the Evening Sky

Index

Acceleration
　gravity, 1–5
　orbits, 42, 43
Algol, 102–6
Analemma, 24
Aristotle, 1, 40, 46, 79

Balmer lines, 117
Big Bang, 153, 158, 159
Binary stars, 131, 132
　eclipsing, 97, 102, 107–14
Blackbody, 72, 143
Brahe, Tycho, 79
Burton, Robert, 6

Cannon, Annie Jump, 117
Celestial equator, 12, 13, 15, 26
Celestial meridian, 12
Celestial pole, 12, 13, 15
Cepheids, 97, 98
Color, 74
Color index, 143
Comte, Auguste, 115
Copernicus, Nicolas, 31, 46

Day
　length, 12
　sidereal, 22
　solar, 22
Declination, 16, 17
Delta Cephei, 97–101, 128, 129
Distance modulus relation, 120, 125, 138, 144
Doppler effect, 59–61, 135, 156, 157

Earth
　age, 67
Eclipsing binaries, 97, 102, 107–14
Ecliptic, 12, 14, 15, 26
Einstein, Albert, 68, 89, 154
Energy levels, 85
Entropy, 165–70
Eotvos, Baron Roland von, 5
Equinox, 16
Eratosthenes, 19
Error estimates. See mean absolute deviation

Focal length, 32
　measurement, 33, 34
Frequency, 59
Friedmann, Alexander, 154
Fusion, 68

Galaxy
　distance measurements, 125, 127, 128, 130, 131–39
　formation, 153

Galileo Galilei, 1, 6
　on Aristotle, 1
　motion, 40, 41
　telescopic observations, 31, 32, 46
Globular clusters, 141, 142
　ages, 147, 148
Goodricke, John, 98, 102
Gravity, 41
　Newton's Law of, 43
Guardian of Forever, 89

Half-life, 56
Harriot, Thomas, 46–48
　sunspot drawings, 52–55
Helmholtz, Herman von, 67
Herschel, William, 97
Hertzsprung-Russell diagram, 121–24, 140–42, 144
Hour
　of right ascension, 17
　Roman, 21
Hubble, Edwin P., 154
Hubble's law, 153–59, 161
Hubble time, 158, 159

Ionization, 88
isotope, 56

Jupiter
　Galileo's observations, 31, 46
　rotation, 61–64

Kirchhoff, Gustav, 115

Latitude, 12, 13
Leavitt, Henrietta Swan, 97, 127, 128
Lens
　eyepiece, 34
　focal length, 32
　objective, 34
Light
　Doppler effect, 59–61, 135, 156, 157
　wave nature vs. particle nature, 89
Light curve, 98, 102, 108
Light gathering power, 35
Lippershey, Hans, 31

Macrostate, 165
Magellanic clouds, 127
Main sequence, 140, 141, 144
Main sequence lifetime, 141
Mass-luminosity relation, 120, 140
Mean absolute deviation (MAD), 3, 8
Microstate, 165
Moon
　Galileo's observations, 31, 46
　parallax, 81–83
　phase nomenclature, 38
　phases, 36–39

Nova, 131, 132
Nova Cygni, 1975, 132–34

Opacity, 91

Parallax, 78
Payne, Cecelia, 117, 118
Pendulum, 4, 6–11
Period
　pendulum, 4
　sidereal, 50
　synodic, 48, 50
Period-luminosity relation, 97, 100, 127, 128
Photon, 86, 90
　number as a function of temperature, 169
　scattering, 93–95
　and transport of energy, 90, 91
Precision, 3

Quasar, 160

Random walk, 95, 96
Red shift, 156, 160–64
Right ascension, 16, 17

Saturn
　Galileo's observations, 31
　rotation, 65–66
Scattering, 93–95
Schmidt, Maarten, 160
Solar spectrum, 73
Specific entropy, 170
Spectrum
　absorption, 115–19
　emission, 86, 116
　thermal (blackbody), 143
Stars
　formation, 140
　masses, 112, 113
　sizes, 108–12
Stefan-Boltzmann law, 72, 143
Sun
　and analemma, 24–28
　and definition of noon, 21
　energy source, 67, 68
　motion in sky, 22
　power output, 67–70
　time over horizon, 18
Sundial, 21, 22
Sunspots, 46–48, 77
　drawings by Harriot, 52–55
Supernova, Tycho's, 79
Systematic errors, 3

Temperature, 71, 72
Turnoff point, 141
Transit, 12

Universe
 age, 158, 159
 expansion, 153, 154

Velocity, 41, 42
 Doppler effect, 59–61, 156, 157
 radial, 60
 wave, 59

Water clock, 21
Wein's law, 72, 143
White dwarf, 131, 132

Young, Thomas, 89

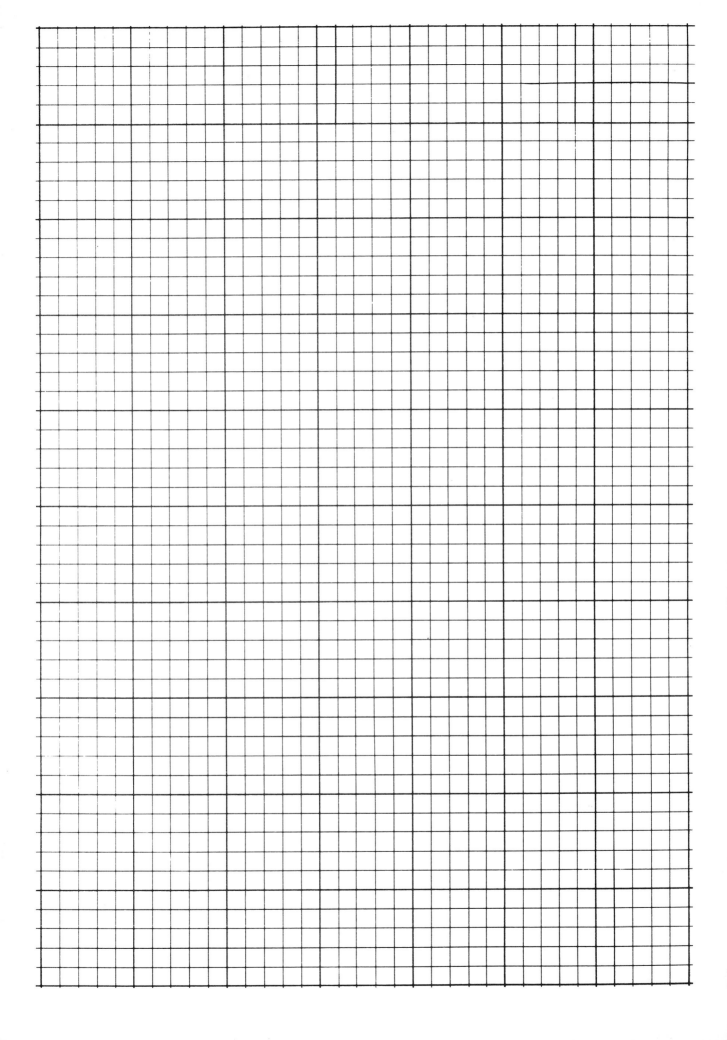

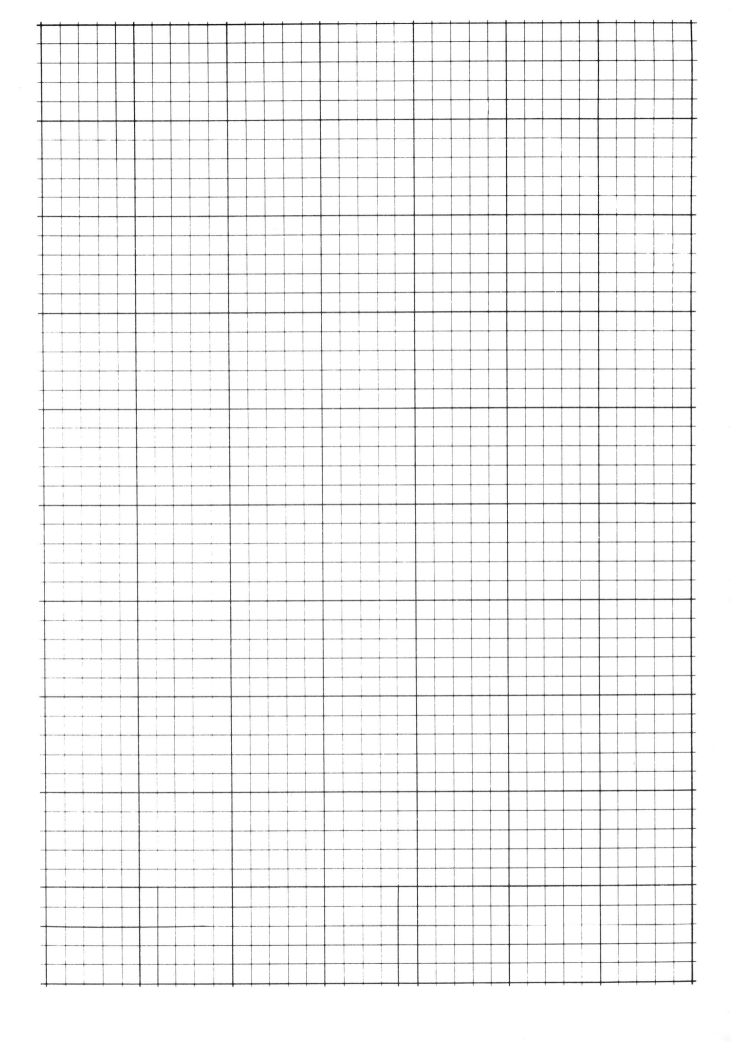

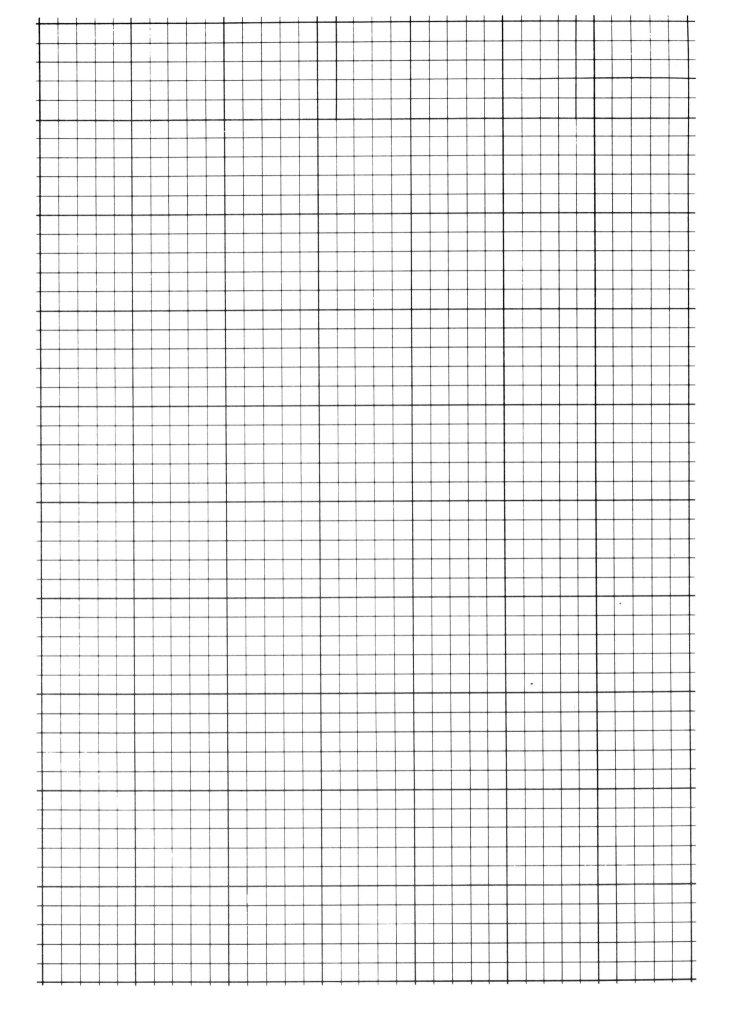

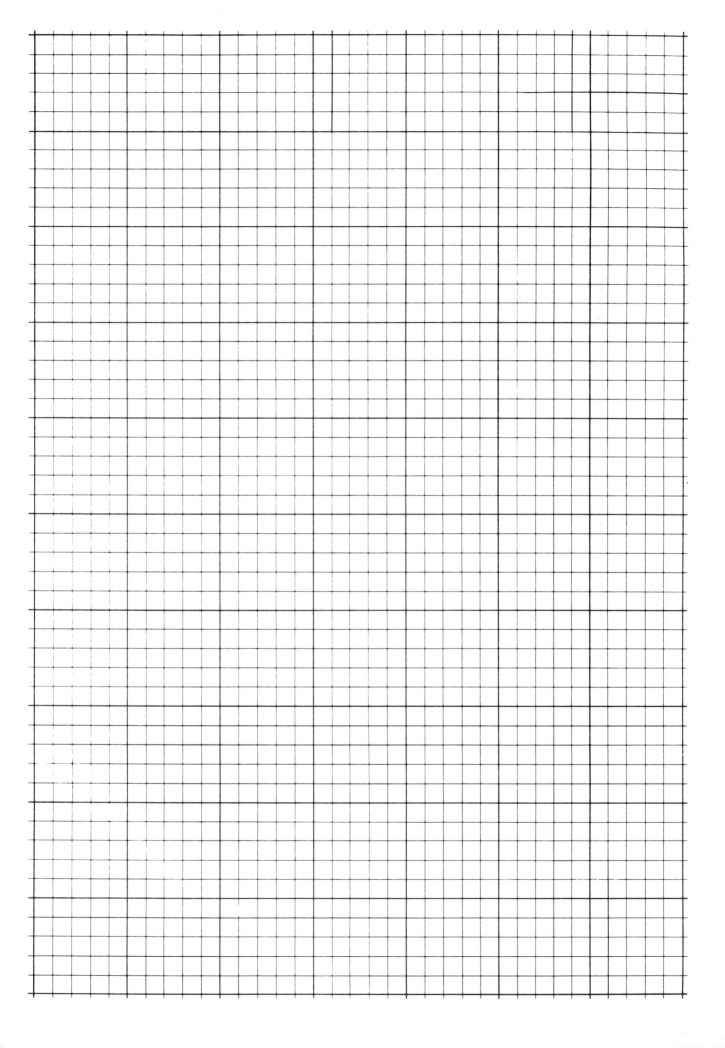

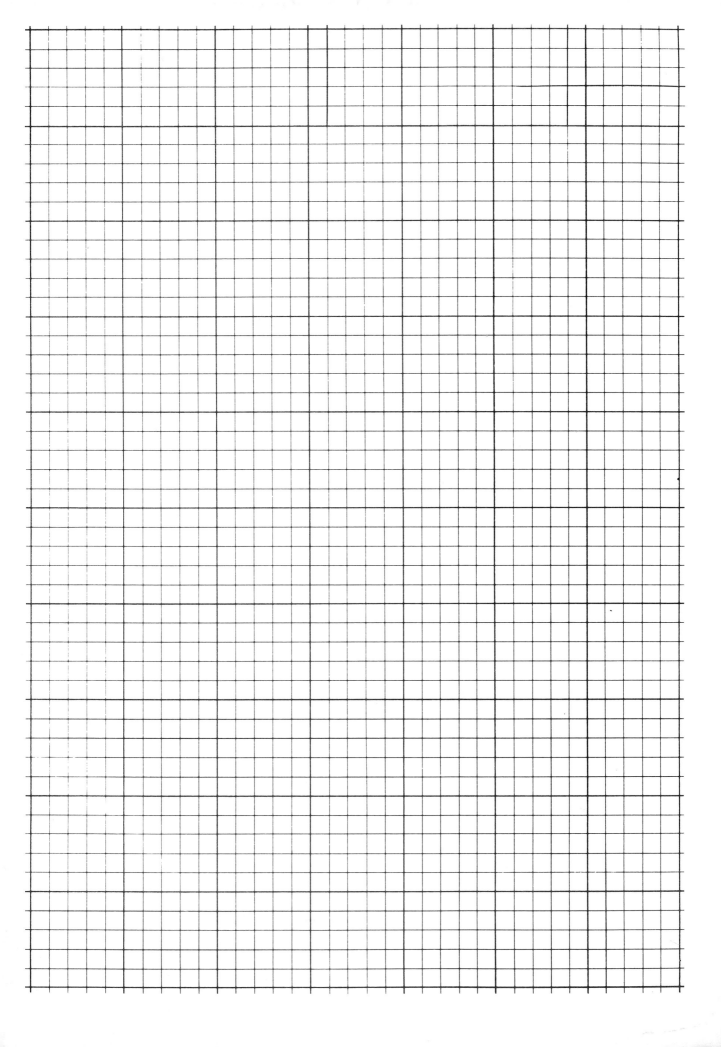